Chemistry for Biologists

Second Edition

The INSTANT NOTES series

Series editor
B.D. Hames
School of Biochemistry and Molecular Biology, University of Leeds, Leeds, UK

Animal Biology
Ecology 2nd edition
Genetics 2nd edition
Microbiology 2nd edition
Chemistry for Biologists 2nd edition
Immunology
Biochemistry 2nd edition
Molecular Biology 2nd edition
Neuroscience
Psychology
Developmental Biology
Plant Biology
Bioinformatics

Forthcoming titles
Cognitive Psychology
Physiological Psychology

The INSTANT NOTES Chemistry series
Consulting editor: Howard Stanbury

Organic Chemistry
Inorganic Chemistry
Physical Chemistry
Medicinal Chemistry
Analytical Chemistry

Chemistry for Biologists

Second Edition

J. Fisher

Department of Chemistry,
University of Leeds, Leeds, UK

and

J.R.P. Arnold

Yorkshire Cancer Research
Laboratory of Drug Design,
University of Bradford, Bradford, UK

BIOS Scientific Publishers
Taylor & Francis Group

LONDON AND NEW YORK

© Garland Science/BIOS Scientific Publishers, 2004

First published 2000
Reprinted 2001, 2002
Second edition published 2004

A CIP catalogue record for this book is available from the British Library.

ISBN 1 85996 355 2

Garland Science/BIOS Scientific Publishers
4 Park Square, Milton Park,
Abingdon, Oxon OX14 4RN, UK and

29 West 35th Street, New York,
NY 10001–2299, USA
World Wide Web home page: www.bios.co.uk

Garland Science/BIOS Scientific Publishers is a member of the Taylor & Francis Group

Distributed in the USA by
Fulfilment Center
Taylor & Francis
10650 Toebben Drive
Independence, KY 41051, USA
Toll Free Tel.: +1 800 634 7064; E-mail: taylorandfrancis@thomsonlearning.com

Distributed in Canada by
Taylor & Francis
74 Rolark Drive
Scarborough, Ontario M1R 4G2, Canada
Toll Free Tel.: +1 877 226 2237; E-mail: tal_fran@istar.ca

Distributed in the rest of the world by
Thomson Publishing Services
Cheriton House
North Way
Andover, Hampshire SP10 5BE, UK
Tel.: +44 (0)1264 332424; E-mail: salesorder.tandf@thomsonpublishingservices.co.uk

Production Editor: Andrea Bosher
Typeset by Phoenix Photosetting, Chatham, Kent, UK
Printed by Biddles Ltd, Guildford, UK, www.biddles.co.uk

CONTENTS

ABBREVIATIONS

ADP	adenosine diphosphate
ATP	adenosine triphosphate
Bq	Becquerel
CD	circular dichroism
CI	chemical ionization
Ci	Curie
d	dextrorotatory (clockwise)
DCCI/DCCD	dicyclohexylcarbodiimide
DMT	dimethoxytrityl
E	entgegen (opposite)
E1	unimolecular elimination
E2	bimolecular elimination
Ea	activation energy
EI	electron impact
EM	electromagnetic
ESI	electrospray ionization
FAB	fast atom bombardment
FAD	flavin adenine dinucleotide
fid	free induction decay
FT	Fourier transform
G	Gibbs free energy
Gy	Gray
HOMO	highest occupied molecular orbital
IR	infrared
IUPAC	International Union of Pure and Applied Chemistry
l	levorotatory (anticlockwise)
LCAO	linear combination of atomic orbitals
LUMO	lowest unoccupied molecular orbital
MO	molecular orbital
MS	mass spectrometry
NAD^+	nicotinamide adenine dinucleotide (oxidized form)
NADH	nicotinamide adenine dinucleotide (reduced form)
NMR	nuclear magnetic resonance
nOe	nuclear Overhauser effect
[O]	oxidizing agents
OH	hydroxyl group
ORD	optical rotatory dispersion
R	rectus (clockwise, chiral center)
S	sinister (anticlockwise, chiral center)
SN1	unimolecular nucleophilic substitution
SN2	bimolecular nucleophilic substitution
TBDMS	tertiary butyl dimethylsilyl
u	atomic mass unit
UV	ultraviolet
Z	zussamen (together)

PREFACE TO SECOND EDITION

In the 3 years that have passed since the publication of the first edition we have received significant feedback on the book's content and style. Happily the majority of this has been very favourable! Amongst the comments we have read have been a few suggestions for modifications or additions to the text. We have taken note of these in preparing this new edition, whilst also incorporating some changes we ourselves felt would improve the readers understanding. A number of the more significant changes are outlined below.

As the book is for biologists and the like, and as water is the biological solvent, we have provided (in Section C) more information about how and why this solvent is so unique. We now include a section on small inorganic molecules of biological interest (Section F) which permits us to cover molecular nitrogen and molecular oxygen, previously not considered except in passing in the section on metals (Section G). In the first edition nucleic acids and proteins were mentioned in terms of their building blocks. We have taken the opportunity with this new edition to include information regarding their respective roles. In doing this we thought it sensible to consider other biological macro-molecules, hence we have included a topic covering the composition and function of lipids (Section K). We have looked again at the section on thermo-dynamics and have made some modifications to our discussion of Gibb's free energy and the statistical definition of entropy. Finally, for completeness we have included a description of circular dichroism in the section on spectroscopy (Section Q). CD is a very useful technique in structural biochemistry. We, in common with a number of reviewers, felt this topic should be presented along-side the spectroscopic techniques more commonly used by chemists.

For Elizabeth and Abigail, as always!

PREFACE TO FIRST EDITION

Students of the biological sciences require a good grounding in chemistry if they are to fully understand key aspects of their courses. Many, but not all, students choosing to study biology, biochemistry and the like, have studied chemistry to 'advanced' level or equivalent. However, the knowledge gained at this stage is generally quickly forgotten, does not usually go far enough or it does not make relevant connections to biological systems. There are of course numerous excellent texts available which cover in great detail many of the topics which impinge on the biological sciences. Some of these attempt to set the material in a biological context. For the most part students studying chemistry as a subsidiary subject often find such texts overwhelming. A number of books have been written on chemistry aimed specifically at biological or life sciences students, and these are a useful source of reference. However, in general we have found that these do not cover the material that we require our biological sciences students to learn. As the number of such students at Leeds is currently over 400 each year we felt it worthwhile writing this book.

Instant Notes in Chemistry for Biologists aims to cover all aspects of chemistry relevant to the biological sciences and is based largely on the contents of the lecture course given by both chemistry and biochemistry departments at the University of Leeds. The book is divided into 15 sections containing 52 topics. Each topic consists of a Key Notes panel with concise statements of the key points covered. These are expanded on in the main part of the topic which includes simple and clear black and white figures which may readily be reproduced for essays or examinations for example. The key notes are essentially a revision aid hence the main body of the text is best read first. The ordering of the topics reflects the fact that some fundamental principles need to be learnt at the outset. Once the early sections have been digested the later topics may be dipped into subsequently. The topics are extensively cross-referenced to assist learning and understanding.

The contents of the book includes aspects of organic, inorganic and physical chemistry. In some instances the division between these branches of chemistry is obvious, in others less so. We have endeavored to cover material in a manner appropriate to the biological sciences and in doing so break down unnecessary divides. Section A provides an introduction to key features of the elements, from atomic structure, to the periodic table, through to isotopes: including natural and synthetic radioisotopes. This leads to Section B and the description of bonding and molecular shape and how these may be represented on paper. In Section C the properties of water and phosphoric acid are discussed in terms of chemistry but as examples of extremely biologically important small molecules. The vast majority of all other biological molecules are carbon based and consequently Section D is aimed at explaining why life should have evolved based on this element.

Section E introduces the concept of isomerism, paying particular attention to stereoisomerism and the labeling of absolute 3-dimensional structures of stereoisomers.

In Section F the emphasis shifts from molecules of importance to the elements, particularly metals of importance in biology.

A key feature of many biochemical polymers is the network of weak inter-actions which hold large structures together, such as hydrogen bonds and hydrophobic interactions. These are discussed in Section G.

In Section H definitions are provided for the various reactive species in chemical reactions. Mechanisms and factors influencing these are included. This leads naturally to Section I in which the key properties of functional groups which frequently occur in biochemistry are discussed. In Section J a similar approach is adopted in considering a special functional group, the aromatic ring. This then leads to section K in which the chemical synthesis of biochemical polymers is described.

Section L covers acid-base properties of aqueous solutions, with a description of buffer solutions and the concept of solubility.

In Section M the fundamentals of thermodynamics are set out. This is a necessary prelude to Section N in which rates of reaction, enzyme kinetics and catalysis in general are discussed.

Finally in Section O the principle of quantitization of energy is introduced together with a description of the electromagnetic spectrum. The information available from the application of various components of this spectrum; ultra-violet, infrared and radiowaves is also covered.

This book is not intended to provide an in depth coverage of chemical concepts, such as books included in the Further Reading list, but rather an overview of key biochemically relevant material. It will work well with other books in this series, in particular *Instant Notes in Biochemistry*.

Julie Fisher, John Arnold

To Elizabeth and Abigail

A1 THE PERIODIC TABLE

Key Notes

Atomic structure

The atom consists of a nucleus containing protons and neutrons and is surrounded by electrons. Electrons are located in specific energy shells and discrete packets, or quanta, of energy are required for electrons to move from one shell to another. The number of protons found at the nucleus is defined as the atomic number.

Quantum mechanics

The idea that electrons resided in specific allowed energy states proved very useful in explaining some physical observations. However, why electrons should be distributed in this way was not explained until the principle of wave-particle duality was developed. The wave-like behavior of electrons is termed quantum mechanics.

Atomic orbitals

The electrons which surround the nucleus of an atom occupy regions of space referred to as orbitals, or subshells. The size of the nucleus and the number of electrons determines the shape and overall energy of a particular orbital. These orbitals are labeled 1s, 2s, 2p, 3p, 4d, 5f, etc. The letter refers to the shape of the orbital, the number to the shell location.

Periodicity

Properties such as valency (number of electrons in the outer shell), and electronegativity (electron-withdrawing or attracting power) of the elements vary in a regular manner according to atomic number. The periodicity of these properties is reflected in the arrangement of rows and columns in the periodic table.

Related topics

Electron configuration (A2)   Molecular orbitals (B1)
Isotopes (A3)   The early transition metals (G1)

Atomic structure

At the turn of the 18th century, the generally held view of the atom was that put forward by John Dalton. His atomic theory had as its basis that atoms were indestructible. It was not until around the turn of the 19th century that this was shown to be incorrect. Atoms are actually comprised of **subatomic particles** known as **neutrons**, **protons**, and **electrons**. The bulk of the mass of the atom is made up of the mass of the neutrons and protons, each given an atomic mass of 1, the mass of the electron being negligible. The neutron has no charge whereas the proton has a single positive charge and the electron a single negative charge:

Hence, the (neutral) atom has the same number of electrons as protons. As the number of protons does not vary for a particular element (see Topic A3) the **number of protons** is the **atomic number**. It was not until the end of the 19th century that the organization of these particles with respect to each other was determined. Experiments involving bombarding metal films with the subatomic particles emitted by various **radioisotopes** (see Topic A3) revealed that the atom had a core or

nucleus where the bulk of its mass was housed; the neutrons, protons and electrons must therefore occupy space surrounding the nucleus.

In 1913 the Danish physicist Niels Bohr postulated that electrons were not simply free to circulate around the nucleus but instead that they must occupy specific allowed energy states; referred to as **energy levels** or **shells**. Discrete packets of energy, or **quanta** (see Topic Q1), are expelled or absorbed to move an electron from one principal shell to another. There is a maximum number of electrons that can occupy each shell (see *Table 1*)

Table 1.   Electron occupancy of principal shells

	Low energy ────────────────▶ High energy				
Energy shell	1	2	3	4	5
Electron capacity	2	8	18	32	32
Number orbitals	s	s, p(x3)	s, p(x3), d(x5)		

Quantum mechanics

Bohr's model of the atom was highly successful, especially in predicting the emission spectrum of the hydrogen atom. However, it provided no underlying reason as to just *why* electrons should be layered in shells within atoms. The answer came from **wave-particle duality**. During the early part of the 20th century, a number of experimental observations were made in which electrons behaved as if they were classical particles, whilst on other occasions they acted as if they were waves. The wave-like character of electrons was evident, for example, in the patterns formed by beams of electrons passing through crystals. This mixture of characteristics has caused philosophical problems ever since, as it became evident during the 1920s that sub-atomic matter had a strange, fuzzy existence; Einstein was not satisfied by this, but was unable to devise a more satisfactory description. In general, a particle can be described by a wave, the amplitude (or strictly, the amplitude^2) of which is related to the probability of finding the particle at a given position. The wave-like behavior is termed **quantum mechanics**, or sometimes **wave mechanics**. The wave is a mathematical function, often called the wave function. This function strongly depends on the physical forces acting on the particle, usually these are electrostatic influences of nuclei and/or electrons. So, when a negatively charged electron is associated with a postively charged proton (as in the hydrogen atom), the electrostatic attraction between the two causes only certain electron wave functions to be allowable. These gave rise to atomic orbitals, which are grouped in shells, just as in the Bohr model.

Atomic orbitals

Within each electron shell there are **subshells**. These subshells have a degree of 'fine structure', in that there are specific locations within the subshell where there is a higher probability of finding an electron than in other regions. These regions are referred to as **atomic orbitals**. These orbitals have unique shapes and can hold two electrons provided these electrons are spinning in opposite directions (see Topic A2). The shapes of the orbitals are specified by the letters 's', 'p', 'd', and 'f' etc., and are shown in *Fig. 1*.

Different energy levels have different numbers of these orbitals as shown in *Table 1*.

Periodicity

Mendeleev in the mid-1800s noted that various properties of the elements seem to go through cycles as atomic number is increased. The boiling points, for example,

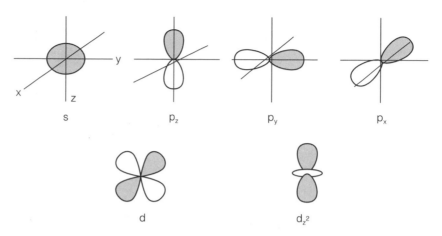

Fig. 1. Shapes of some of the atomic orbitals. There are three 'p' orbitals in the x,y,z directions and five 'd' orbitals in each shell where these appear. Orbitals show boundaries within which 95% probability of finding electron.

do not increase with atomic number but go through peaks and troughs (see *Fig. 2*).

The energy required to remove an electron (the ionization energy) followed a similar trend (see *Fig. 3*)

Also it was noted that the number of bonds that an element could form with another element varied with atomic number (see Topic B2 and *Table 2*).

An atom of an element with atomic number 3, 11 or 19, is able to combine with one atom of some other element, and is thus referred to as **monovalent**. Those with atomic numbers 5 and 13 are able to form three bonds and are referred to as being **trivalent** elements, and so on. It should be noted that elements with atomic

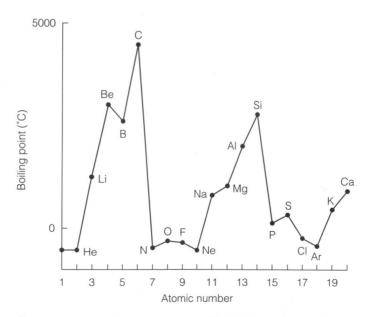

Fig. 2. Variation of boiling point with atomic number.

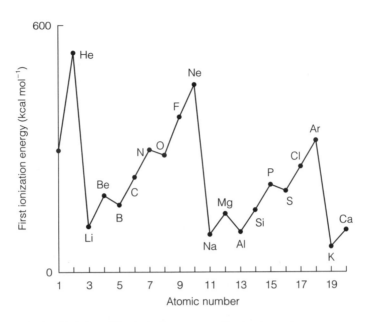

Fig. 3.   Variation of first ionization energy with atomic number.

numbers 2, 10 and 18 etc. do not readily form bonds with other elements. This is due to the fact that these elements have a complete outer shell of electrons which makes them very stable. Indeed these elements are called the **inert** or **noble** gases.

These trends are of course all reflected in the modern organization of the periodic table (see *Fig. 4*).

The periodic table is arranged in columns. In each column the elements have the same outer shell **electronic configuration** (see Topic A2) and therefore the **same valence**. As illustrated in *Fig. 3*, elements in Group I require the input of a relatively small amount of energy to lose an electron, i.e. to ionize. This readiness to lose an electron has led to these and the Group II elements being termed **electropositive**.

In contrast, it is extremely difficult to remove an electron from the Group VII elements; these are close to the stable complete outer shell position and therefore have a stronger tendency to acquire electrons. Consequently these elements are referred to as being **electronegative**.

It is the ability of an element to lose or accept electrons that dictates its chemistry (see Topic B2).

Table 2.   Variation of bond capacity with atomic number

Atomic number		No. of bonds	Example	
(a)	(b)		(a)	(b)
3,	11	1	LiH	NaH
4,	12	2	BeH_2	MgH_2
5,	13	3	BH_3	AlH_3
6,	14	4	CH_4	SiH_4
7,	15	3	NH_3	PH_3
8,	16	2	OH_2	SH_2
9,	17	1	FH	ClH
10,	18	0	–	–

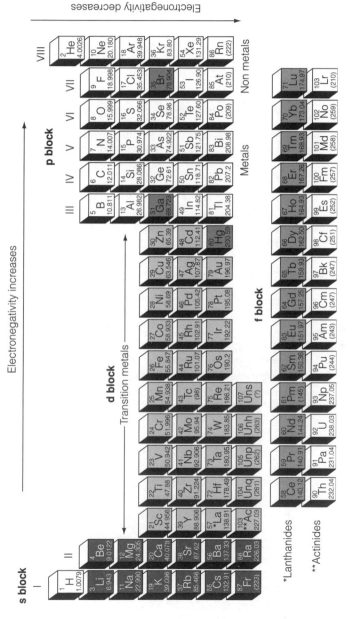

Fig. 4. The periodic table of the elements. The shading groups elements with similar properties.

A2 ELECTRON CONFIGURATION

Key Notes

Aufbau principle	The word 'aufbau' is German for 'building up'. The aufbau principle states that the lowest energy orbital will be the first filled with electrons.
Pauli exclusion principle	The Pauli exclusion principle states that only two electrons may be placed in each orbital and these electrons must have opposed spins.
Hund's rule	Hund's rule states that if two or more empty orbitals of equal energy are available, one electron must be placed in each orbital until they are all half-filled.
Abbreviated configuration	A shorthand method for describing the electronic configuration of an atom has been developed. For example, $1s^1$ and $1s^2$, signify that 1 and 2 electrons, respectively, are present in the 1s orbital.
Related topics	Molecular orbitals (B1) Nature of chemical bonding (B2)

Aufbau principle

The arrangement of electrons in atomic or molecular orbitals (see Topic B1) is known as the **electronic configuration**. In determining the electronic configuration a number of rules have to be considered. The first is that the **lowest energy orbital must be filled first**. Therefore, all atoms have electrons in the 1s level, and so on. This is the known as the **aufbau principle**.

Pauli exclusion principle

The Pauli exclusion principle states that each orbital has a maximum occupancy of two electrons. An electron has a property called a **spin**, and can be thought of as rotating about an axis. Of course it can rotate from west to east or east to west, therefore it has two possible orientations, or two spins. These are denoted as an up ($\uparrow$) and down ($\downarrow$) arrow. The Pauli exclusion principle states that two electrons in the same orbital must have opposite spins (see *Fig. 1*).

Hund's rule

Each electron shell is at a different energy level. Within each shell there are the **subshells** called **atomic orbitals** (see Topic A1). From shell 2 onwards, more than one orbital type is present. In level 2 there is one 's' and three 'p' orbitals. The 'p' orbitals are all at the same energy level, (the 's' orbital slightly lower) and are thus referred to as being **degenerate**. Hund's rule says that if there are two or more degenerate orbitals then these should be singly occupied until they are all half filled, with spins aligned.

Using the aufbau, Pauli and Hund rules it is possible to write the electron configurations of the ground state, the lowest energy state, of all the elements (see *Fig. 2*).

	1s	2s	2p	2p	2p
H	↑				
He	↑↓				
Li	↑↓	↑			
Be	↑↓	↑↓			
B	↑↓	↑↓	↑		

Fig. 1. Organization of electrons in atomic orbitals according to the Pauli exclusion principle.

	1s	2s	2p	2p	2p
C	↑↓	↑↓	↑	↑	
N	↑↓	↑↓	↑	↑	↑
O	↑↓	↑↓	↑↓	↑	↑
F	↑↓	↑↓	↑↓	↑↓	↑
Ne	↑↓	↑↓	↑↓	↑↓	↑↓

Fig. 2. Organization of electrons in atomic orbitals following the Pauli exclusion principle and Hund's rule.

Abbreviated configuration

The format adopted in *Fig.* 2 for electronic configurations is somewhat cumbersome. It is more convenient to 'abbreviate' this picture as shown in *Fig.* 3. Clearly the configuration for boron ($1s^2 2s^2 2p^1$) signifies 2 electrons in the 1s orbital, 2 in the 2s and 1 in the p orbitals. A further abbreviation is possible by referring to the previous inert gas. Thus, the electronic configuration of lithium is, with reference to helium, (He) $2s^1$ (see *Fig.* 3).

H	$1s^1$		
He	$1s^2$		
Li	$1s^2$	$2s^1$	
Be	$1s^2$	$2s^2$	
B	$1s^2$	$2s^2$	$2p^1$
C	$1s^2$	$2s^2$	$2p^2$
Li	(He)	$2s^1$ (i.e. the configuration of helium plus a 2s electron)	
Na	(Ne)	$2s^1$ (i.e. the configuration of neon plus a 2s electron)	

Fig. 3.   Some abbreviated electronic configurations.

A3 ISOTOPES

Key Notes

Definition	Atoms which have the same number of protons but differing numbers of neutrons are referred to as isotopes of each other.
The mole	One mole of any element or compound has the identical number of formula units as atoms that are present in 12.0000 g of carbon-12.
Stable isotopes	Stable isotopes are those which under normal conditions do not transform into other element types. They do not require special handling.
Radioisotopes	Radioisotopes are isotopes that decompose and in doing so emit harmful particles and/or radiation. There are naturally occurring radioisotopes (e.g. ^{238}Ur) and those that may be prepared synthetically (e.g. ^{3}H, ^{32}P).
α-emitters	Isotopes which emit a particle consisting of 2 protons and 2 neutrons (the nucleus of Helium) are referred to as α-emitters and the unit of mass number 4 is an α-particle.
β-emitters	Isotopes whose nuclei can lose an electron are referred to as β-emitters and the stream of electrons emitted from the nucleus are referred to as β-particles.
γ-emitters	Isotopes which emit a photon of energy in addition to losing particles of β- or α-radiation are referred to as γ-emitters. As the energy has no mass this is referred to as γ-radiation rather than γ-particles.
Half-life	The half-life is the time required for the activity of a radioisotope to have decayed to 50% of the original activity.
Units and measurement	The unit of activity of radiation is the Becquerel (Bq), and the unit of dose is the Gray (Gy). Film dosimeters, Geiger counters, and scintillation counters are all used to measure radiation.
Radiation damage and sickness	Alpha and β-particles, and X-rays and γ-rays cause the formation of unstable ions or free radicals when they pass through the body. These highly active species can modify the cells genetic material and lead to a range of symptoms; from nausea to cancers.
Use of radioisotopes	Radioisotopes are used in the food industry for prolonging product life. In biological sciences they are used as an analytical tool, and in medicine as a diagnostic tool.
Related topics	The periodic table (A1) Electron configuration (A2)

Definition   The number of protons for any given element is unique. That is, all carbon atoms have only 6 protons (this is the **atomic number**), however the number of neutrons is not unique. Most atoms of naturally occurring carbon contain 6 neutrons; these atoms have a **mass number** of 12. However, around 1.1% of all carbon atoms have 7 neutrons, and therefore have a mass number of 13. Thus, atoms of an element that have the same number of protons but different numbers of neutrons are **isotopes** of that element. Isotopes of a particular element undergo identical chemical reactions.

The mole   The most abundant isotope of carbon, carbon-12, is used as the basis for the SI definition of Avagadro's number, or the **mole**. One mole of a compound has the identical number of mass components as atoms in 12 g of carbon-12. As the actual mass of atoms in grams is very small a convention has been adopted for expressing atomic mass in terms of atomic mass units; one **atomic mass unit**, 1u = 1.665620×10^{-24} g. Taking account of the 1.1% natural abundance of carbon-13, the atomic mass of carbon may therefore be expressed as 12.011 u (see Topic A2).

Stable isotopes   Many elements have isotopes that are stable under normal conditions. They do not emit ionizing radiation or otherwise spontaneously decay. For example, hydrogen has two natural isotopes, both of which are stable; **hydrogen** (or protium) and **deuterium**. Hydrogen has one proton but no neutrons, deuterium has one proton and one neutron. To differentiate between the two one would normally write;

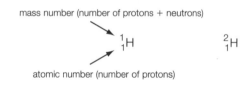

Hydrogen   **Deuterium**

(Deuterium is a special case and is often written simply as 'D'.)
Hence, the two stable isotopes of carbon would be written as follows;

In some cases the natural abundance of stable isotopes can be more equivalent. For example, bromine has two stable isotopes, bromine-79 and bromine-81, with natural abundances of approximately 50% each. Therefore the atomic mass of bromine is approximately 80 u.

Radioisotopes   Isotopes of a number of elements are unstable and will readily lose mass or energy in order to occupy a stable state; these are naturally occurring **radioisotopes** or **radionuclides**. In all instances, the mass or energy loss can cause damage to the surrounding environment, the extent of the damage depends on the nature of the loss. In general, naturally occurring radioisotopes can emit three forms of radiation; α-particles, β-particles and γ-radiation (or rays). After the decay event an entirely different element remains; this process is referred to as **transmutation**.

Other radioisotopes may be prepared synthetically *via* the bombardment of otherwise stable atoms with subatomic particles. The first non-natural transmutation, observed by Rutherford, was the conversion of nitrogen-14 to oxygen-17 by α-particle bombardment. The process was found to proceed *via* the formation of a **compound nucleus**;

$$\underset{\text{Alpha particle}}{{}^{4}_{2}\text{He}} \quad + \quad {}^{14}_{7}\text{N} \longrightarrow \underset{\text{Compounded nucleus}}{{}^{18}_{9}\text{F}} \longrightarrow {}^{17}_{8}\text{O} \quad + \quad \underset{\text{Proton}}{{}^{1}_{1}\text{p}}$$

Also molybdenum-99 is produced from the bombardment of molybdenum-98 by neutrons;

$$ {}^{98}_{42}\text{Mo} \quad + \quad {}^{1}_{0}\text{n} \longrightarrow {}^{99}_{42}\text{Mo} \quad + \quad {}^{0}_{0}\gamma$$

This is one of the most important isotopes used in diagnostic medicine.

α-emitters

Alpha particles are clusters of 2 protons and 2 neutrons, that is the nuclei of helium atoms, and as such are the largest and most highly charged of particles given off in radioactive decay. These particles move quite slowly, around one-tenth the velocity of light, and are susceptible to collisions with surrounding elements. During a collision an alpha particle will lose its energy and charge. Owing to their size and velocity, α-particles are not generally considered hazardous as they cannot penetrate the outer layer of skin. However they can cause a severe burn. One isotope of uranium is an α-emitter, uranium-238, and is transmuted in to an isotope of thorium;

$$ {}^{238}_{92}\text{U} \longrightarrow {}^{236}_{90}\text{Th} \quad + \quad {}^{4}_{2}\text{He, } \alpha$$

As thorium-236 is itself unstable, further decay processes take place (see *Fig. 1*).

Table 1 shows other examples of α-emitters.

β-emitters

Beta particles are actually electrons, however they are produced within the nucleus. They have, of course, a smaller charge than α-particles (-1, compared with $+2$) and are much lighter, therefore they can travel rapidly. Beta particles can be emitted with a range of energies. At the low energy end β-particles cannot penetrate the skin, high energy β-particles can penetrate skin and indeed reach internal organs.

To emit an electron from the nucleus a neutron must change into a proton. In doing so, essentially no mass is lost but the atomic number increases by 1 unit;

$$ \underset{\substack{\text{Tritium}\\\text{Radioactive isotope of hydrogen}}}{{}^{3}_{1}\text{H}} \longrightarrow {}^{3}_{2}\text{He} \quad + \quad \underset{\text{Beta particle}}{{}^{0}_{-1}\text{e}}$$

Tritium is transmutated to helium-3.

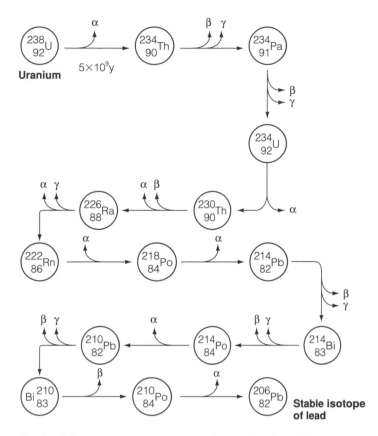

Fig. 1. *Schematic representation of the decay of Uranium-238.*

γ-emitters

Gamma radiation often accompanies the breakdown of natural radioisotopes by the emission of α- and β-particles. Gamma rays are **high energy photons** which escape at the same time as the radionuclide is moving to a more stable state. Gamma rays have no mass and no charge and therefore are highly penetrative and highly dangerous; akin to X-rays (of which they can be thought as being high energy versions).

$$^{234}_{90}\text{Th} \longrightarrow {}^{234}_{91}\text{Pa} + {}^{0}_{-1}\text{e} + {}^{0}_{0}\gamma$$

Half-life

Since by definition radionuclides are unstable, over a period of time they decay. The rate at which decay takes place varies from element to element. The rate of decay is not usually noted. Rather the time taken for a given sample of a pure iso-tope to decay to 50% of its original activity, that is the half-life is referred to. According to the information provided in *Table 1*, after 800 years 100 g of radium-226 will have reduced to 50 g of radium-226. In another 800 years, only 25 g would be left. The radioactive decay is exponential.

Units and measurement

The SI unit of radioactivity is the **Becquerel** (Bq), and is the number of **disintegra-tions per second** (dps). The older unit of activity is the **Curie** (Ci), 1 Ci being equivalent to 3.7×10^{10} Bq. A sample with an activity of 1 Ci would be extremely active. Hence when considering acceptable radiation levels, one talks in terms of

Table 1. Some natural and synthetic radioisotopes

Element	Isotope	Half-life	Radiation
Natural isotopes			
Uranium	$^{238}_{92}U$	4.5×10^9 years	alpha
Thorium	$^{230}_{90}Th$	8.0×10^4 years	alpha, gamma
Radium	$^{226}_{88}Ra$	1600 years	alpha, gamma
Radon	$^{222}_{86}Rn$	3.8 days	alpha, gamma
Rhenium	$^{187}_{75}Re$	7.0×10^{10} years	beta
Samarium	$^{149}_{62}Sm$	4×10^{14} years	alpha
Potassium	$^{40}_{19}K$	1.3×10^9 years	beta, gamma
Synthetic isotopes			
Strontium	$^{90}_{38}Sr$	28.1 years	beta
Cesium	$^{137}_{55}Cs$	30 years	beta
Iodine	$^{131}_{53}I$	8.0 days	beta
Phosphorus	$^{32}_{15}P$	14.3 days	beta
Oxygen	$^{15}_{8}O$	124 sec	positron
Tritium	$^{3}_{1}H$	12.26 years	beta
Carbon	$^{14}_{6}C$	5730 years	beta

milli Ci (mCi or 10^{-3}Ci) and so on. This unit is still in use in the health service and research laboratories.

The unit of **absorbed radiation** is the **Gray** (Gy) and this is defined as 1 joule of energy per kilogram of tissue. The older unit of dose was the **rad**, which is one-hundredth of a Gray.

There are several devices which may be used to measure radiation levels. First, a **dosimeter**, as the name suggests, is used for measuring exposure. Dosimetry usually takes the form of a film badge. The photographic film is fogged by radiation and the degree of fogging relates directly to the dose exposed to.

A **Geiger counter** is used to measure activity. It consists of an evacuated tube fitted with two electrodes and into which a gas is introduced at very low pressure. Beta and γ-particles entering the tube have sufficient energy to ionize the gas. The ions thus formed are counted.

Scintillation counters rely on the fact that certain substances, such as phosphorus, emit spark-like flashes of light when hit by ionizing radiation; this can be translated in to doses of radiation received, or radiation counts (see *Fig. 2*).

Radiation damage and sickness

Atomic radiation is dangerous as it can lead to the formation of ions or **free radicals** within a living system. It does this by knocking off electrons from molecules that it strikes (see *Fig. 3*). The ions and free radicals formed can then react with

100% of disintegrations can never be detected, actual disintegrations per minute is given by:

$$dpm = cpm \times 100/percentage\ efficiency$$

typical efficiencies of scintillation counters are:

^{3}H 35% $\qquad\qquad$ ^{14}C 80%

If ^{3}H phenylalanine, with specific activity 1 Ci mol^{-1}, is used in synthesis of a protein, and a scintillation counter measures 4500 cpm from 1 mg of protein, the number of moles of phenylalanine incorporated in the protein is determined as follows:

(a) $\qquad$ 4500 cpm × 100/35 $\qquad$ = 12 857 dpm
$\qquad$ (actual disintegrations)
(b) $\qquad$ 12 857 dpm × 1/60 $\qquad$ = 214 dps
$\qquad\qquad\qquad\qquad\qquad\qquad\qquad$ = 2.14 × 10^2 Bq
$\qquad$ (conversion to dps)
(c) $\qquad$ 2.14 × 10^2 × 1/(3.7 × 10^{10}) = 5.8 × 10^{-9} Ci
$\qquad$ (conversion to Ci)
(d) $\qquad$ specific activity $\qquad\qquad$ = 1 Ci mol^{-1}
$\qquad$ 5.8 × 10^{-9} moles of ^{3}H labeled phenylalanine are incorporated per milligram of
$\qquad$ protein.

Fig. 2. Example of use of radioactivity counting to determine level of incorporation of a radiolabeled amino acid to a protein.

other molecules. If these reactions take place in the cell then the result could be genetic mutations and subsequent tumor growth. High doses of radiation cause cell death.

The first symptoms of exposure to radiation are detected within cells that are growing and dividing most rapidly, such as in the bone marrow; the white blood cell count drops dramatically. Nausea, dehydration, hemorrhaging, and hair loss are some of the early signs of radiation exposure.

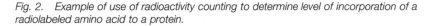

Hydroxyl radical

Fig. 3. Bombardment of water by atomic radiation leads to a proton and an hydroxyl radical.

Uses of radioisotopes

Radiation treatment of foodstuffs

In some major countries approval has been granted for ionizing radiation to be used in the food industry to prolong the shelf life of perishable goods. A low dose of up to 1000 Gy can effectively eradicate any insects which remain following harvest and reduces the tendency of potatoes to sprout during storage.

A stronger dose, perhaps up 10 000 Gy, is required to significantly reduce the amount of salmonella bacteria in meat and fish. Similarly this is required to stave-off the formation of molds on soft fruits such as raspberries and strawberries. However, the expense of this exercise and the likely public concern at seeing food with the marking 'radiochemically protected' means that this is not a common practice to date.

Analytical tool in biological sciences
In the biological sciences, radionuclides are frequently used as an analytical tool. The radioactive isotope of phosphorus, ^{32}P, is routinely used to analyze the sequences of DNA and RNA molecules. The ^{32}P label may be introduced to the 3' or 5' end of the oligonucleotide chain (see Topic M2) *via* the use of **phosphoryl transfer catalyzing enzymes**, from a donor (^{32}P labeled) molecule. The DNA or RNA is then cleaved at specific sites using another enzyme and the resultant mixture is analyzed by polyacrylamide gel electrophoresis. An autoradiogram of the gel is then produced, on which bands are detected for fragments with the ^{32}**P tag**. The sensitivity of the photographic plate to this β-emitter means that only very small amounts of material are required for the analysis.

Carbon-14, with a half-life of 5730 years, is utilized to trace metabolic pathways and to 'age' ancient material.

Radioisotopes in medical diagnosis
Radiation has been used as a medical diagnostic tool for over a century. The use of X-rays to examine bone and hard tissue is well understood, but other radionuclides are used for other applications. Technetium-99 is the most common nuclide used in medicine. It is a breakdown product of ^{99}Mo (see *Fig. 4*)

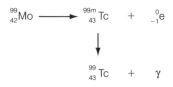

Fig. 4. *The radioactive decay of molybdenum-99 (m, metastable).*

Cobalt-60 was the radiation source traditionally used to treat cancers until the end of the 1950s. It has a half-life of 5.3 years and emits β- and γ-radiation. It is a very active source, much more powerful gram for gram than radium, for example.

Iodine-123 was used to monitor the performance of the thyroid gland, this is the only use of iodine in the body. The patient would simply drink a glass of water containing iodide-123 ions and, with an appropriate detector placed on the skin near the thyroid gland, the γ-rays given off by this radionuclide can be detected. Iodine-123 is useful due to its short half-life. However, this has been superseded by the use of a technetium salt which can just as readily penetrate the thyroid, but has a much shorter half-life, 6 h, making it safer to use. Iodine-131 has a half-life similar to ^{123}I. In addition to γ-rays it emits very damaging β-particles. Consequently, ^{131}I would generally only be used to treat cancers of the thyroid.

Other radionuclides routinely used in medicine include gallium-67 (^{67}Ga). It has a half-life of 78 h and is a γ-emitter used in, for example, the diagnosis of Hodgkin's disease.

B1 MOLECULAR ORBITALS

Key Notes

Definition	A molecular orbital is formed by the combining of atomic orbitals from two atoms of a (covalent) bond. When two 's' atomic orbitals combine two molecular orbitals are formed; a bonding molecular orbital (σ) containing the bonding electrons and an antibonding molecular orbital (σ^*) which is of higher energy and, under normal conditions, unoccupied.
Hybridization	The combining of atomic orbitals within an atom to form new orbitals is termed hybridization. Bonds involving hybrid orbitals are more stable than those involving 'pure' atomic orbitals.
sp^3	The four hybrid orbitals resulting from the mixing of three 'p' orbitals and one 's' orbital are sp^3 orbitals. The four new orbital lobes point to the corners of a tetrahedron.
sp^2	The three hybrid orbitals resulting from the mixing of two 'p' orbitals and one 's' orbital are sp^2 orbitals. The three new orbital lobes point to the corners of an equilateral triangle.
sp	The two hybrid orbitals resulting from the mixing of one 'p' and one 's' orbital are sp orbitals. The two new orbital lobes are 180° apart (linear).
Related topics	The periodic table (A1) Electron configuration (A2)

Definition

When atoms come together to form a **covalent bond** (see Topic B2) the outer shell electrons of the atoms interact, and the space that they are most likely to occupy changes; this is the **molecular orbital**.

For example, when two hydrogen atoms combine to form a hydrogen molecule, H_2, their **spherical 1s** orbitals combine to form an **elliptical** shaped **molecular orbital**. This molecular orbital is at a lower energy level than the isolated s-orbitals and hence this is the orbital which will be filled with those electrons involved in bonding (see *Fig. 1*). The interaction between atomic orbitals may be represented on an **energy level diagram** (see *Fig. 2*). Each bonding interaction results in a bonding molecular orbital (σ) and, a higher energy, antibonding molecular orbital (σ^*).

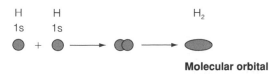

Molecular orbital

Fig. 1. The combination of spherical 1s-orbitals of two hydrogen atoms to form an elliptical molecular orbital.

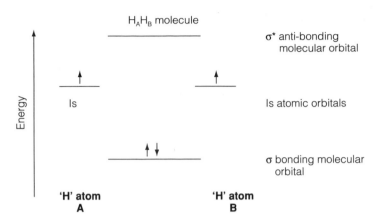

Fig. 2.   Energy level diagram for the interaction between two 1s-orbitals.

If nonbonding electrons are present then these occupy a nonbonding molecular orbital (n) which is intermediate in energy.

When orbitals overlap along a bond axis, the **covalent bond** (see Topic B2) formed is referred to as a **sigma** (σ) bond. However, if the orbital overlap is above or below the bond axis, as is possible when all but s-orbitals are involved, then **pi** (π) bonds are formed (see *Fig. 3*).

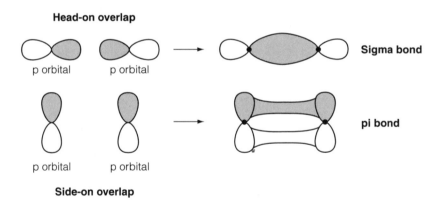

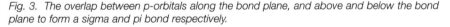

Fig. 3.   The overlap between p-orbitals along the bond plane, and above and below the bond plane to form a sigma and pi bond respectively.

Hybridization

The structures of many small compounds cannot be explained by invoking a simple orbital overlap picture for orbitals that are singly occupied in the electronic ground state of an atom. However, Linus Pauling showed that a mathematical combination of s- and p-orbitals results in **hybrid orbitals** which are then involved in bonds that are more stable than their nonhybridized counterparts (see *Fig. 4*). Hybrid orbitals are also found involving s-, p-, and d-atomic orbitals (see Topic A1).

sp³

Carbon is tetravalent (see Topics A1 and B2). It has four electrons in its outer shell, two of which are paired in the 2s-orbital, while the other two are unpaired in the 2p-orbitals. This would suggest that carbon could form compounds such as

Fig. 4. An sp-hybrid orbital formed by the combination of an s- and a p-orbital.

CH_2 (see *Fig. 5a*). This molecule does exist but only fleetingly as it is extremely unstable. Therefore, carbon must adopt an electronic configuration (see Topic A2) which is different to that of the ground state. This alternative state must have one of the '2s' electrons promoted into the vacant 'p' orbital; resulting in the electronic configuration $1s^2 2s^1 2p_x^1 2p_y^1 2p_z^1$. This electronic configuration would suggest that four bonds could form but of two different types. However, there is evidence to suggest that all CH bonds in, for example, methane are identical; hence, the orbitals involved in bonding must be identical. Linus Pauling showed the mathematical result of mixing three 'p' orbitals with one 's' orbital; four new sp^3 orbitals with lobes at the corners of a tetrahedron (see *Fig. 5b*).

(a)

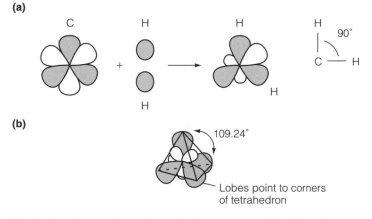

(b)

Fig. 5. The 3-dimensional shape of carbon compounds in (a) the absence and (b) the presence of hybrid orbitals.

The advantage of an s- and p-hybrid orbital over an s- or p-orbital lies in the asymmetry of the orbital. One of the lobes of the sp-orbital is much larger than the other. This larger lobe can overlap more efficiently with the lobe of any incoming atomic orbital forming a more stable sigma bond (see *Fig. 6*)

sp^2

The most common electronic state of carbon is sp^3 but it is not the only state. If the 2s-orbital was combined with only two of the 2p-orbitals, the result would be an **sp^2 hybrid** with one unhybridized 2p-orbital remaining unchanged. The three sp^2 orbitals are arranged in a plane. When two sp^2 carbons approach each other they form a strong sigma bond by sp^2–sp^2 overlap. When this occurs, the unhybridized

Fig. 6. Overlap between an sp-hybrid orbital and any other orbital is more efficient than similar interactions involving only s- or p-orbitals.

p-orbitals overlap to form a pi (π) bond. The combination results in the net sharing of four electrons and the formation of a carbon-carbon double bond. To complete the structure, four hydrogens are required to form sigma bonds with the four remaining sp^2 orbitals (see *Fig. 7*).

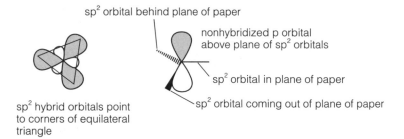

Fig. 7.   The sp^2 hybrid orbitals of carbon in ethene.

sp

If only one of the p-orbitals is combined with an s-orbital the result is **two sp-hybrid** orbitals. These are linear or 180° apart. The remaining p-orbitals are unaffected but are involved in pi-bonding when a similarly hybridized atom approaches (see *Fig. 8*).

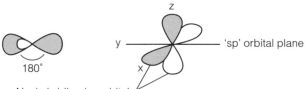

Fig. 8.   The sp-hybrid orbital of carbon in ethyne.

B2 NATURE OF CHEMICAL BONDING

Key Notes

The octet rule	Atoms of elements tend to undergo those chemical reactions which most directly result in them gaining the electronic configuration of the nearest noble gas. When atoms are in the first row of the periodic table, this stable configuration requires eight electrons to fill the valence shell. Hence, atoms are said to undergo reactions which obey the octet rule.
Ionic bonds	An ionic bond is formed between atoms that are negatively charged and those that are positively charged. Such charged atoms, or ions, are attracted to each other and stabilize each other but do not transfer or share electrons between each other. Thus, this is more of an interaction than a bond as there is no extra build up of electron density at the mid-point between the two atom centers.
Covalent bonds	A covalent bond is formed between elements that in general are not readily ionizable; they are neither strongly electropositive nor electronegative. Atoms involved in covalent bonds share the bonding electrons and thus attain an apparent complete valence shell. In covalent bonds electron density is distributed between the atom centers.
Lewis structures	The Lewis structure of a molecule depicts the electrons in the valence shell of atoms involved in bonding. Each electron is depicted as a dot and this is a straightforward way of keeping check of electron balance in a molecule or reaction.
Molecular orbital theory	The molecular orbital theory is an approach to describe covalent bonding in molecules, by considering the combination of atomic orbitals. Bonding, antibonding and nonbonding molecular orbitals are introduced and, following on from this, the shapes of simple molecules may readily be described.
Coordination compounds	Many elements are known to have fixed valences, however a number of elements can exhibit two or three stable valences. These can readily be explained by consideration of the relevant atomic and molecular orbitals. A number of metals have combining powers that cannot readily be reconciled in terms of covalent or ionic bonding. Such metal-containing compounds are now referred to as coordination complexes, as groups or ligands around the central metal donate a lone pair of electrons to unoccupied orbitals of the metal.
Related topics	The periodic table (A1) The early transition metals (G1)

The octet rule   Atoms whose outer (valence) electron shells hold a complete set of electrons are substantially more stable than those that do not. Consequently, helium and neon are extremely stable, indeed they are inert (see Topic A1). Similarly if by gaining or losing electrons an atom can acquire a complete valence shell, the stability of the ionized atom is such that electron loss or gain will readily take place. For example, the electronic configuration of sodium is (Ne) $3s^1$ (see Topic A2). Consequently if sodium were to lose one electron it would have the electronic configuration of the stable element neon ($1s^2\,2s^2\,2p^6$). That such an ionization can occur is supported by the relatively low ionization energy for sodium (see Topic A1). Similarly, chlorine has the configuration (Ne) $3s^2\,2p^5$. In gaining one electron it would obtain the electronic configuration of argon. Hence, Cl^- is relatively readily formed. For most of the lighter elements a complete outer or valance shell requires eight electrons. Consequently the observations referred to above are said to obey the **octet rule**.

It should be noted that for atoms with available d-orbitals, the valence shell can expand beyond the octet (see Topic G1).

Ionic bonds

Ions of opposite charges are strongly attracted to each other. When two ions, for example Na^+ and Cl^- (as examples of ions of elements at opposite extremes of the periodic table), come together an **ionic compound** is formed which is an orderly array, or **lattice**, of oppositely charged atoms. The attraction between these atoms is referred to as the **ionic** bond; as electrons are not shared between the oppositely charged ions an ionic bond is really an interaction rather than a bond. Ionic bonds within lattices are very strong, and therefore ionic compounds have very **high melting points**. As ionic compounds are charged they are generally good **conductors of electricity**, and are soluble in polar media such as water.

Covalent bonds

A covalent bond is formed when atoms come together to share an electron pair. A covalent bond will form when an ionic bond is not favorable, that is, when neither atom is readily ionizable. Consequently covalent bonds generally involve atoms of the inner columns of the periodic table (see Topic A1). For example, two nitrogen atoms can form a covalent bond with each other by the sharing of three valence electrons. Thus, for each atom there is an apparent complete octet. To form any covalent bond the atomic orbitals of the bonding atoms must overlap in order that electrons may be shared, and in doing so **molecular orbitals** are formed. In the case of N_2 these orbitals are referred to as σ (sigma) and π (pi) bonding molecular orbitals. Electrons occupy molecular orbitals obeying the same rules as guide atomic orbital occupation (see Topics A1 and A2) and the positions of the orbitals once occupied are such that electron repulsions are minimized.

Covalent compounds are usually gases, liquids, or low melting point solids. Their generally neutral nature means that they are usually insoluble in polar solvents such as water.

Lewis structures

A simple shorthand method for indicating covalent bonds in molecules is to use **Lewis** structures. In these the valence electrons of an atom are shown as dots (see *Fig. 1*)

For the lighter elements, a stable molecule results whenever a noble gas configuration with filled s- and p-orbitals is achieved (see Topic A1).

Molecular orbital theory

The **molecular orbital** (MO) theory was developed to help describe covalent bonds. In MO theory the prime assumption is that if two nuclei are positioned at some equilibrium distance and their valence electrons are added, these electrons will be 'accommodated' in molecular orbitals. Of the various methods of

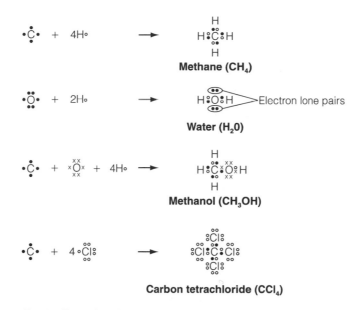

Fig. 1.   Examples of some Lewis structures for a number of small molecules.

approximating molecular orbitals, their shapes and properties, the most straight forward is a **linear combination of atomic orbitals** (LCAO). Thus, the combination of two *s*-atomic orbitals results in two molecular orbitals labeled σ and σ* for the bonding and antibonding molecular orbitals respectively (see Topic B1). In the bonding molecular orbital the atomic orbital overlap is positive and the electron density between the atoms is increased, this is a low energy situation. In the antibonding MO, the overlap is negative and the electron density is decreased, this is a high energy situation (see *Fig. 2*). These molecular orbitals are idealized as they depict the situation when similar atoms are bound. More realistically when bonds form between heteroatoms then the molecular orbital picture is somewhat distorted. Using carbon monoxide as an example, oxygen is more electronegative than carbon, thus the most stable arrangement for bonding electrons is for them to be located near to the oxygen atom. The bonding molecular orbital thus resembles the atomic orbitals of oxygen and the antibonding those of carbon (the less electronegative element) (see *Fig. 3*).

Coordination compounds

The compounds formed by a number of the heavier metals cannot be rationalized in terms of concepts of ionic or covalent bonding, where single electrons are gained, lost or shared (see Topic G1). For example, platinum exhibits valences of +2 and +4, and yet chlorides of this element will react with ammonia (in which the valence of nitrogen and hydrogen are already satisfied) to form the compound $PtCl_2 \cdot 4NH_3$. This and similar molecules could be rationalized, at least in part, by assuming that 'ligands' (the 'valence satisfied groups') donated a lone pair of electrons to occupy vacant orbitals of the coordinating metal (see *Fig. 4*), thus forming **coordinate** or **dative** bonds.

Coordination compounds occur in a number of biological systems (see Topics G2–G4).

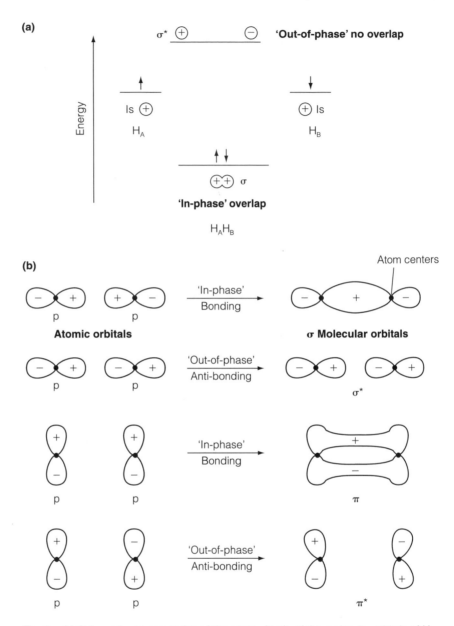

Fig. 2.   (a) Schematic representation of the energy levels of the molecular orbitals of H_2, and (b) shapes of linearly combined atomic orbitals, both with positive and negative overlap (+, − = phase of orbital).

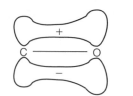

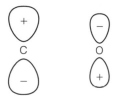

π shape distorted in favor
of more electronegative
atom, oxygen

π* shape distorted in favor
of less electronegative
atom, carbon

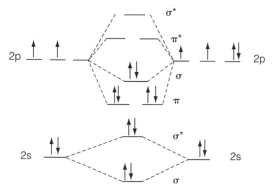

Fig. 3.   Energy level diagram depicting the covalent bonds formed between carbon and oxygen to form carbon monoxide.

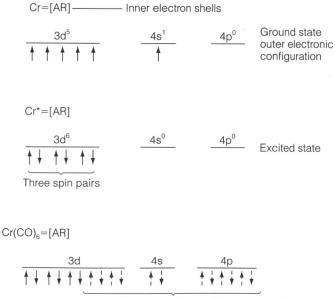

Fig. 4.   Schematic representation of electron coordination between chromium and carbon monoxide to form Cr(CO)₆.

B3 SHAPES OF SOME SMALL MOLECULES

Key Notes

Methane

In methane (CH_4) the carbon atom is sp^3 hybridized. The four sp^3 orbitals point to the corners of a regular tetrahedron where overlap with four hydrogen s-orbitals results in the formation of four σ and σ* molecular orbitals. The H-C-H bond angle is 109.5°, as in tetrahedral geometry.

Ammonia

The nitrogen atom of ammonia (NH_3) has five electrons in its valence shell, two in the 2s-orbital and three in the 2p-orbitals. Hybridization results in four sp^3 hybridized orbitals one of which is doubly occupied (this is a lone pair). The four orbitals point to the corners of a tetrahedron but only three of these orbitals are available for bonding. Hence, three σ, three σ*, and one nonbonding molecular orbitals are formed with three hydrogen atoms. The H-N-H bond angle is 107.1°.

Water

In water the oxygen atom is sp^3 hybridized. In this case two of the four atomic orbitals are doubly occupied and all four point to the corners of a tetrahedron. The valence of oxygen is satisfied by the bonding of two hydrogen atoms. The H-O-H bond angle is 104.5°.

Ethene

In ethene (CH_2CH_2) the two carbon atoms are sp^2 hybridized. These three orbitals point to the corners of an equilateral triangle. A fourth, p-orbital, which is singly occupied, is perpendicular to the plane of the triangle and is available for orbital overlap (and hence bonding) which would result in electron density above and below the plane of the carbon-carbon bond (a pi-bond).

Formaldehyde

In formaldehyde (CH_2O), or methanal, both the carbon and oxygen atoms are sp^2 hybridized. Two of the sp^2 orbitals on the oxygen atom are doubly occupied. These and the third point towards the corners of an equilateral triangle. Perpendicular to the plane of the triangle is a singly occupied p-orbital available for overlap with a similar orbital on the carbon atom.

Related topic

Molecular orbitals (B1)

Methane

In methane the carbon atom is sp^3 hybridized (see Topic B1). Thus four equivalent atomic orbitals, arranged to point to the corners of a regular **tetrahedron**, are available for bonding (see *Fig. 1*). Upon overlap with the s-orbital of four hydrogen atoms, four σ (bonding) and four σ* (antibonding) molecular orbitals are formed. As the four C-H bonds are equivalent all of the H-C-H bond angles are equivalent at 109.5°.

Since the valence of each of the bonding atoms is satisfied this molecule is relatively inert.

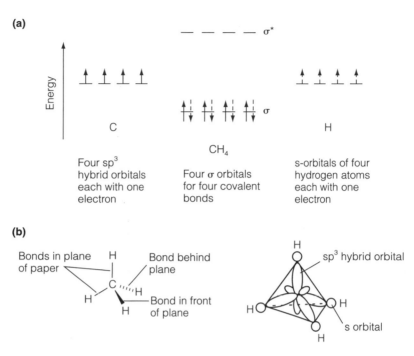

Fig. 1.   The shape of methane. (a) The molecular orbital picture and (b) the 3-dimensional arrangement of atoms.

Ammonia

The nitrogen atom of ammonia is also sp^3 hybridized, hence four atomic orbitals are available. However, prior to bonding one of these orbitals is occupied by two electrons. The repulsion between this and a singly occupied orbital is greater than that between two singly occupied orbitals. Consequently a slightly **elongated tetrahedral** arrangement of these atoms results. When these atomic orbitals overlap with the s-orbital of three hydrogens, the resultant H-N-H bond angle is compressed to 107.1° (see *Fig. 2*).

Although in forming ammonia the valences of both hydrogen and nitrogen are satisfied, the presence of a pair of electrons 'owned' solely by the nitrogen (**the lone pair**) makes the nitrogen a relatively active centre. Ammonia, and similar nitrogen compounds, can thus act as **bases** (see Topic N1) and **nucleophiles** (see Topic I1).

Water

The central oxygen atom of water is sp^3 hybridized, with two of these orbitals being doubly occupied with electrons (two lone pairs) belonging to oxygen. Consequently a water molecule adopts a similar structure to that of ammonia and methane. However, the tetrahedron is elongated to an even greater extent, due to the repulsion between two lone pairs, and the H-O-H bond angle is compressed to 104.5° (see *Fig. 3*).

As a result of two lone pairs of electrons, like ammonia the oxygen of water can be considered as a reactive center. Indeed, water can also act as a **base** (see Topic N1) or a **nucleophile** (see Topic I1).

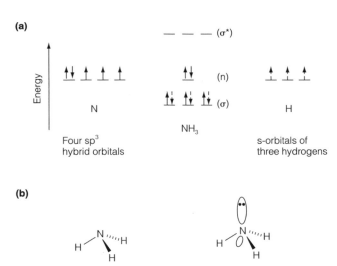

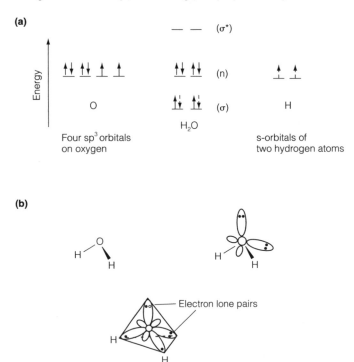

Fig. 2.   The shape of ammonia. (a) The molecular orbital picture and (b) the 3-dimensional arrangement of atoms [n, nonbonding (lone pair) electrons].

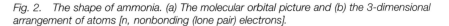

Fig. 3.   The shape of water. (a) The molecular orbital picture and (b) the 3-dimensional arrangement of atoms.

Ethene

The two carbon centers of ethene are each sp^2 hybridized (see Topic B1). Consequently, to minimize repulsion between the electrons in each of these, the orbitals point to the corners of an **equilateral triangle**. The remaining singly occupied p-orbital lies perpendicular to the plane of the triangle and thus when involved in bonding, it results in electron density being remote from the C-C atom axis (see *Fig. 4*). This electron density can be thought of as being relatively 'exposed' and consequently makes ethene a relatively reactive compound, susceptible to attack by **electrophiles** (see Topics I1 and I2).

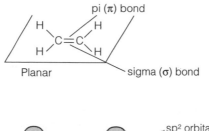

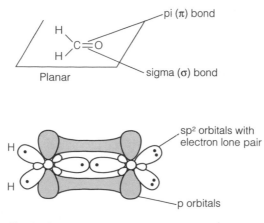

Fig. 4. The shape of ethene.

Formaldehyde

In formaldehyde both the carbon and oxygen atoms are sp^2 hybridized (see *Fig. 5*). Again as oxygen has two lone pairs of electrons it is able to act as a **base** (see Topic N1). As oxygen is more **electronegative** than carbon (see Topic A1) the carbon-oxygen bond is **polarized** (see Topics I3 and J2) such that the carbon becomes slightly positively charged. Consequently the carbon center is susceptible to **nucleophilic attack** (see Topic I1).

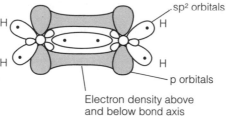

Fig. 5. The shape of formaldehyde.

B4 DRAWING CHEMICAL STRUCTURES

Key Notes

Chemical formulae

There are several ways in which the composition of a compound may be described. Each way differs in the actual detail of the information provided. It is possible to describe a molecule by its empirical, molecular, or structural chemical formulae. The empirical formula is simply the ratio of the various elements in a compound, the molecular formula is the actual number of each element in the compound, and the structural formula indicates the bonding arrangement amongst the elements.

Molecular structures

Information about the structure of a molecule may be conveyed in a number of ways. The most obvious is in terms of a structural formula in which all atoms are indicated in a 2-dimensional fashion. The tetrahedral nature of carbon, for example, may be indicated by means of solid, dashed, or wedged lines.

Related topics

The periodic table (A1)
Molecular orbitals (B1)

Naming organic compounds (D2)

Chemical formulae

A molecule may be defined at different levels. The simplest description requires that the ratios of the elements in the molecule be specified. This is the **empirical formula**. Hence, for example, the empirical formula for cyclohexane and cyclopentane is CH_2, and for benzene and ethyne (acetylene is its nonsystematic name) it is CH. This is the information that would be obtained from an **elemental analysis**, but clearly this is not enough to identify a compound uniquely. A more precise description is provided by the **molecular formula** which is the actual

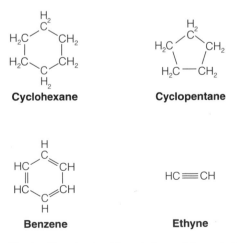

Fig. 1.   The structural formula for cyclohexane, cyclopentane, ethyne and benzene. A line represents two electrons of a sigma (σ) or pi (π) bond.

number of each element in a molecule; for cyclohexane C_6H_{12}, cyclopentane C_5H_{10}, and for benzene C_6H_6 and ethyne C_2H_2.

The most detailed description of a molecule is provided by the **structural formula**, in which the bonding relationship between atoms is specified. Hence, the structural formulae for the four molecules referred to are as shown in *Fig. 1*.

A better example is provided by two molecules with the same molecular formula C_4H_{10} (see *Fig. 2*)

n-Butane and 3-methylpropane are different compounds and are in fact **structural isomers** (see Topic E1).

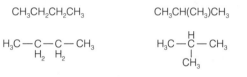

$CH_3CH_2CH_2CH_3$	$CH_3CH(CH_3)CH_3$

n-butane   2-methylpropane

Fig. 2.   *Structural formulae for C_4H_{10}. These molecules are related as structural isomers.*

Molecular structures

Representing molecules other than simply by name can become quite cumbersome, even for relatively small molecules, if all of the atoms in the molecule need to be specified. Consequently a shorthand system for representing chemical formulae has been developed. This system is based around carbon, in that carbon and its associated hydrogens (satisfying valence requirements) may be represented by the vertex of lines meeting: each line representing a bond, or by the end of a line (see *Fig. 3*).

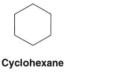

Cyclohexane   **Cyclopentane**

Benzene   **Ethyne**

Fig. 3.   *Abbreviated structural formulae for cyclohexane, cyclopentane, benzene, and ethyne.*

It should be noted that hydrogens attached to atoms other than carbon must be specified (see *Fig. 4*) as must all other atoms.

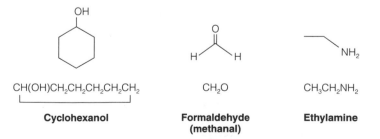

$CH(OH)CH_2CH_2CH_2CH_2CH_2$

Cyclohexanol CH_2O $CH_3CH_2NH_2$

          Formaldehyde   **Ethylamine**
          (methanal)

Fig. 4.   *Examples of abbreviated structural formula for molecules with elements in addition to carbon and hydrogen.*

Ultimately it may be necessary to convey the 3-dimensional shape of a molecule in the 2-dimensional space provided on a sheet of paper. This is achieved by the use of single lines in the plane of the paper for one or two of the bonds and dashed lines and wedges to indicate bonds behind the plane and in front of the plane of the page respectively (see *Fig. 5*)

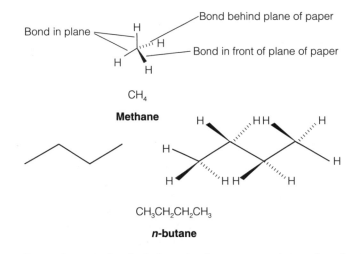

CH_4

Methane

$CH_3CH_2CH_2CH_3$

n-butane

Fig. 5.   Representing the 3-dimensional arrangement of atoms in molecules.

B5 CLEAVAGE OF CHEMICAL BONDS

Key Notes

The curly arrow	Atoms in molecules are held together by the sharing of pairs of electrons. Hence, in forming or breaking a bond, electrons must be relocated. The movement of electrons within a reaction scheme is depicted by the use of a curly arrow. A curly arrow with a full (called double) head is used to indicate the movement of a pair of electrons. A half or single headed arrow is used to indicate the movement of a single electron. In each case the electrons move to the site indicated by the head of the arrow, from the site (atom) at the base of the arrow.
Heterolytic cleavage	When a covalent bond is broken, one of the attached elements may take both of the bonding electrons. The bond is said to have broken in a heterolytic manner. The result of such bond breakage is oppositely charged atoms. This is by far the most common form of bond cleavage under normal experimental conditions.
Homolytic cleavage	If, when a covalent bond is broken, atoms at each end of the bond take an equal share of the bonding electrons (i.e. one each), then homolytic cleavage is said to have occurred. The result of such a process is the formation of two (uncharged) radical species.
Related topics	Nature of chemical bonding (B2) Reactive species (I1)

The curly arrow   In chemical reactions, bonds are broken and formed. As bonds are comprised of electron pairs, electrons are relocated during the course of a reaction. To indicate the movement of electrons within a reaction scheme a **curly arrow** is used (see *Fig. 1*). The arrow indicates where electrons are moving to, and from. Two types of curly arrow are used, one in which the arrow head is full (referred to as a double headed arrow) and one in which the arrow head is halved (referred to as a single headed arrow (see *Fig. 1*)) As the names suggest, the double headed arrow

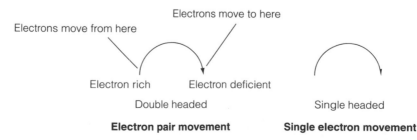

Electrons move to here

Electrons move from here

Electron rich   Electron deficient

Double headed   Single headed

Electron pair movement   **Single electron movement**

Fig. 1.   The use of curly arrows to indicate electron movement.

indicates the movement of an electron pair, the single headed arrow the movement of one electron.

It is important to note therefore that an arrow head will point to a nominally positive (or electron deficient) atom and the arrow base from a nominally negative (or electron rich) atom or bond (see *Fig. 1*)

Heterolytic cleavage

A covalent bond is formed by the sharing of two electrons. When such a bond is broken during the course of a reaction it may break in one of two ways. The first and most common cleavage mechanism results in the bonding electron pair moving to one of the bonding atoms (see *Fig. 2*) resulting in the formation of a negative charge at that atom and a positive charge at the electron-depleted atom. The different natures of the products of this reaction have lead to this process being called **heterolytic cleavage**.

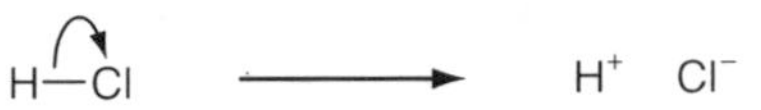

H—Cl $\longrightarrow$ H$^+$ Cl$^-$

Fig. 2.   The movement of electrons in heterolytic bond cleavage.

Homolytic cleavage

The alternative to heterolytic cleavage occurs when the bonding electrons are shared equally by the products of a cleavage reaction. Such a reaction is referred to as **homolytic** as the products each carry a single electron, i.e. they are **radicals** (see *Fig. 3*). Such bond cleavages only occur under special conditions as the products (radicals) are extremely unstable. However, many key events in biology involve the formation of radicals.

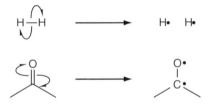

Fig. 3.   The movement of electrons in homolytic bond cleavage.

C1 PHYSICAL CHARACTERISTICS

Key Notes

Physical properties

Water is a liquid over a wide temperature range which is unusual for low molecular weight compounds. Each water molecule is capable of hydrogen bonding to four other water molecules and this explains its temperature behavior. Water has a dipole moment and this contributes to the fact that water is an excellent solvent for charged or polar species.

Structure of water

The structure of liquid water is not well understood, however many structures have been determined for solid water, i.e. ice. These structures differ in the way in which tetrahedral substructures, containing four oxygen atoms, are packed around each other. In ice Ih the tetrahedral units are arranged in a hexagon.

Water for life

Water is involved in almost all physiological processes. The average adult human consists of 50–65% water. Around 70% of the Earth's surface is water and, so far, ours is the only planet known to sustain life. Water appears to be essential for life.

Related topics

Shapes of some small molecules (B3)
Chemical properties (C2)

Hydrogen bonds (H1)

Physical properties

At normal pressure, water is a liquid over an unusually wide temperature range. Its freezing (0°C) and boiling (100°C) define the Celsius temperature scale (which used to be termed Centigrade). These melting and boiling points are distinctly high for a compound of such a low molecular weight. This is most obvious when comparisons are made to related compounds, as in *Table 1*. The origin of this apparent anomaly is connected to the nature of the water molecule itself.

As the water molecule contains two lone pairs of electrons on the oxygen atom, and two polarized hydrogen atoms (see below), each water molecule is able to participate in *four* hydrogen bonding interactions (section H1). It can act as an acceptor for two hydrogen bonds, via the lone pairs, and as a donator for two others, through its hydrogen atoms. This matching of donor and acceptor capacity means that water molecules are optimally suited to hydrogen bonding with

Table 1.   Properties of various compounds analogous to water

Compound	Formula	Molecular weight	Melting point (°C)	Boiling point (°C)
Hydrogen fluoride	HF	20	−83	19.5
Water	H_2O	18	0	100
Ammonia	H_3N	17	−77.7	−33.4
Methane	H_4C	16	−183	−161
Hydrogen sulphide	H_2S	34	−85.5	−60.7

each other. This is what occurs in liquid (and solid) water. These bonds have to break before water can enter the gas phase.

Water, H_2O, is made up of atoms which differ significantly in their electronegativity (see Topic A1). This means that the electrons of the O-H bond are not shared equally between the two atoms; there is greater electron residency at oxygen resulting in a partial negative charge there. This is balanced by a partial positive charge, due to effective electron depletion, at the hydrogen atom. The O-H bond is said to be polarized (see Topic I1), which is often represented thus:

$$O - H$$
$$\delta^- \quad \delta^+$$

As illustrated in Topic B3, the water molecule can be described as having two hydrogen atoms attached to one face of the oxygen atom. This means that the centers of partial negative and positive charge, arising from the polarizations of the two O-H bonds, do not coincide. This is reflected in the fact that the water molecule has a substantial **dipole moment**.

The apparently exceptional melting and boiling points of water, as shown in *Table 1*, can now be rationalized. In their pure states, ammonia and hydrogen fluoride can only form two hydrogen bonds per molecule, rather than four as in water. There is no dipole moment in the methane molecule, and no possibility of hydrogen bonding. Hydrogen bonding in H_2S is very weak. These factors mean that these compounds are far more volatile than water.

Of great importance to biology is the fact that water is an excellent solvent. A wide range of compounds, including many salts, dissolve in water. For salts, being ionic compounds (section B2), this means that the positive and negative ions have to be prised apart. So, for example, sodium chloride dissolves in water even though the forces holding its crystal structure together are very strong (dry NaCl melts at 801°C). This is possible because water is able to **solvate** ions. A positive ion can be surrounded by water molecules, bound via their (partially) negatively oxygen atoms. Likewise, a negative ion is surrounded via the hydrogen atoms of water molecules (see *Fig. 1*). These favorable interactions enable many salts to dissolve. In this way, in liquid water positive and negative ions are effectively shielded from each other, which can also be expressed by stating that water has a relatively high **dielectric constant**.

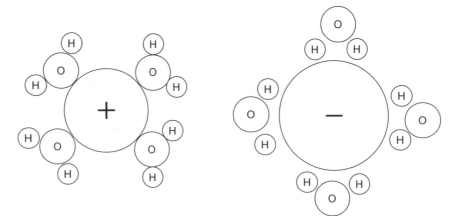

Fig. 1. Solvated ions.

Structure of water

On freezing, water becomes ice. In the structure of ice, each oxygen atom is surrounded tetrahedrally by four other oxygen atoms, with hydrogen atoms in between. In fact there are about 11 different forms of ice, which differ mostly in the packing of these tetrahedral units, which changes in response to applied pressure. However, since pressures are not generally extreme on Earth, or at least where we generally find ice, the only form found naturally is a variety denoted Ice Ih. The 'h' (hexagonal) denotes the type of symmetry of the structure, which is manifested in snow flakes (which always have six corners). The structure of ice Ih is somewhat 'open', with the molecules being less crowded together than when they are jostling against each other in the liquid state (see *Fig. 2*). This means that the ice is less dense than the liquid. Ice therefore floats, and as a result lakes and oceans do not freeze from the bottom up. This is another feature of water which is of global importance.

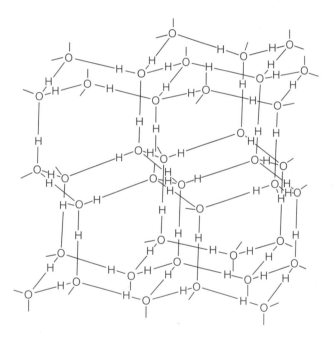

Fig. 2. The structure of Ice Ih.

The structure of *liquid* water is less well understood. A number of models have been developed, all of which involve transient hydrogen bonding networks. In some models these networks can be viewed as momentary aggregates resembling lumps of ice. Other models involve smaller clusters of water molecules. Whatever the case, the high level of hydrogen bonding means that the liquid state is maintained to a higher temperature than would otherwise be expected.

Water for life

Water is generally considered to be essential for life. It participates in almost every physiological process. It is an essential component of active, living systems. The average adult human consists of 50–65% water, the proportion is higher in children. According to current theories, life first developed in the seas, before invading dry land. This ancestral link with the sea is suggested by the similarity of salt levels in the oceans with those in blood.

Whilst water is crucial for biochemical and physiological processes, it is also important to life in a broader, environmental context. About 70% of the Earth's surface is covered by water. The evaporation, condensation, freezing and thawing of water are amongst the most important factors driving weather and climate. In fact, water is the major 'greenhouse gas' in the atmosphere, as it absorbs infra-red radiation (see Topic Q4), warming the atmosphere in the process. In contrast to other, perhaps more familiar, greenhouse culprits like methane and carbon dioxide, human activities seem to have little effect on the level of water in the atmosphere, which comes mostly from evaporation from the oceans.

The presence of water, particularly *liquid* water, is considered to be a vital factor in the possible development of life beyond this planet. So far, liquid water has not been detected on the surface of any other world, apart from the Earth. However, for some as-yet undetermined portion of its history, the planet Mars had liquid water flowing across its surface. Did this provide an opportunity for life to develop, before conditions changed to the cold, dry, airless Martian deserts of today? Where did the water go? At the time of writing, results from orbiting spacecraft suggest that there are significant quantities of subsurface ice on Mars. Perhaps there are underground niches for life to exist at the present time on Mars.

Europa, a moon of Jupiter, is strongly suspected of harboring liquid water beneath a frozen outer shell. The liquid layer is prevented from freezing by heat generated within this moon by tidal forces. This has led to the suggestion that dark oceans of Europa contain life. Just as might be the case on Mars, this might not be highly developed or advanced, but even the discovery of the most primitive cells (or their fossils) would have the most profound philosophical and cultural implications. Whatever the eventual outcome, it is the presence of water which has been the vital pointer.

C2 CHEMICAL PROPERTIES

Key Notes

Reactivity

Water can react with a range of molecule or atom types. It can react with metal oxides to form bases and with non-metals to form acids. It can also form 'association' compounds called hydrates, a number of which have medical applications.

Suspensions and colloids

The term suspension describes a sample of an insoluble substance dispersed in a liquid. In general the mixture is opaque and the insoluble substance may be removed by filtration. A number of medications are administered as water- or oil-based suspensions. Colloids are related to suspensions in that they are liquid samples containing a suspension of tiny particles. These particles are generally so small that they can pass through a filter paper, but sufficiently large to scatter light. Colloids have very special properties. For example they have high adsorption ability and they generally carry an electrical charge, making them very important in biology and medicine.

Emulsions

An emulsion is a mixture of two immiscible liquids, which when shaken vigorously forms a suspension with one of the liquids suspended in the other. In temporary emulsions the suspended droplets would eventually come together and two layers, for the two immiscible liquids, would form. Permanent emulsions may be prepared if the droplets can be prevented from recombining. This is routinely achieved by making the droplets charged.

Related topics   Reactive species (I1)   Aqueous behavior (Section N)

Reactivity

In discussing the chemical properties of water it is first important to note that this is a very stable molecule. Even at very high temperatures, up to around 1600°C, very little decomposition is detected.

When describing organic chemistry, water is simply talked of in terms of its **acid**, **base** (see Section N), or **nucleophilic** properties (see Topics B3 and I1). However many important reactions occur between water and inorganic compounds. Water can react with metal oxides, non-metal oxides and active metals to form inorganic bases, inorganic acids, and to release molecular hydrogen, respectively (see *Fig. 1*).

Water can also form '**association**' compounds. For example when a solution of barium chloride is heated, barium chloride crystals are formed which have associated with them two water molecules per barium atom/ion. Water-associated compounds are referred to as **hydrates**. When the water of hydration is removed from a hydrate, an anhydrous compound is formed. For example:

$$BaCl_2 \cdot 2H_2O \longrightarrow BaCl_2 + 2H_2O_{(g)}$$

Barium chloride dihydrate   Anhydrous barium chloride

Metals

$$2Na + 2H_2O \longrightarrow 2NaOH + H_{2(g)}$$
Sodium hydroxide

Metal oxides

$$CaO + H_2O \longrightarrow Ca(OH)_2$$
Calcium hydroxide

Nonmetal oxides

$$SO_3 + H_2O \longrightarrow H_2SO_4$$
Sulfur trioxide   Sulfuric acid

Fig. 1.   The reactions of water with metals, metal oxides and non-metal oxides.

Some substances lose their hydration waters on exposure to air; these are termed **efflorescent**, others gain water and are said to be **hygroscopic**. A hydrate of importance in the medical field is Plaster of Paris;

$$(CaSO_4)_2 \cdot H_2O \quad + \quad 3H_2O \longrightarrow 2(CaSO_4 \cdot 2H_2O)$$
Plaster of Paris (soft)   Gypsum (hard)

Another hydrate, $MgSO_4.7H_2O$, is commonly used to treat gastric problems and is called **Epsom salts**.

Suspensions and colloids

In discussing processes which occur in biology; natural biochemical processes or those induced by the administration of a drug for example, it is tempting to talk in terms of aqueous **solutions**. However, in many instances this is a misconception. Depending on the 'solute' nature and size the term 'solution' may not be adequate. A solution, by definition contains a solvent, and solute particles that are so small they cannot be observed directly (unless they convey a color on the solution) or removed by filtration; generally this means individual solute molecules dispersed in a 'sea' of solvent molecules. When these conditions do not hold new labels are required to describe the solute/solvent mix; these are **suspensions** or **colloids**. A suspension differs from a solution as it consists of an insoluble substance dispersed in a liquid. As the insoluble substance is dispersed and not dissolved, suspensions are generally **opaque,** solutions are transparent. As soon as a suspension is formed the insoluble matter will gradually clump together either at the top or bottom of the container. Consequently the composition of the suspension is said to be **heterogeneous**. A solution by definition is **homogeneous.** The insoluble matter is generally of such a size that it will not pass through either a filter paper or a membrane.

Many medications are supplied as water suspensions; e.g. milk of magnesia, but others are supplied as oil suspensions; e.g. procaine penicillin G.

Nebulizers depend on the suspension of water droplets in air; a **mist**, to deliver water particles to the bronchial tract.

Colloidal solutions are closely related to suspensions as they consist of tiny particles suspended in a liquid. However they have characteristics which differ sufficiently from suspensions for colloidal solutions to be treated as a separate class.

Colloidal particles range in size from 1 to 100 nm, however their presence can be detected by shining a light beam through the mixture; the particles are large enough to scatter light (the **Tyndall effect**). As they are so small, a given volume of such particles presents a huge surface area upon which smaller molecules may be **adsorbed**. Several commonly used medications are based on this property; e.g. Kaolin is a water-based colloidal solution of finely divided aluminum silicate used for the relief of diarrhea.

Almost all colloidal particles carry a **charge**. This arises when colloid particles selectively adsorb ions. As like charged colloidal particles repel each other they cannot come together to form larger particles that would tend to settle, as in suspensions. Such charged particles can only be forced to move together under the influence of an electric current.

Emulsions

If two immiscible liquids such as water and oil are poured together two layers are formed. Vigorous shaking of the mixture results in the formation of tiny droplets of oil suspended in the water; this is an **emulsion**. Eventually the oil droplets recombine to form an oil layer, therefore the emulsion has been only temporary. A **permanent emulsion** may be prepared if the suspended droplets could be prevented from recombining. Routinely this is achieved by the addition of an **emulsifying agent**; this is a protective charged colloid which coats the surface of the suspended droplets. As the droplets then carry, for example, a positive charge, they cannot recombine due to charge repulsion. An example of a permanent emulsion is mayonnaise. This is a mixture of oil and vinegar with egg yolk as the emulsifying agent. Soap acts as an emulsifying agent on grease and oils in water.

D1 PROPERTIES OF CARBON

Key Notes

Occurrence of carbon	Carbon is the basis for life on this planet. It is a component of every organism. It is the principal component of our daily diet, fuel and clothing. It is perhaps not surprising, therefore, that over 6 million carbon based (hence organic) compounds have been identified compared to around 250 000 inorganic compounds.
Bonding characteristics of carbon	Carbon is near the middle of the second row of the periodic table and therefore has intermediate electronegativity. There are two consequences of this. First, as carbon is neither strongly electropositive, nor strongly electronegative, it readily forms covalent bonds. Moreover, it can form covalent bonds with elements either side of it in the periodic table, resulting in the possibility of huge numbers of compounds, and of course can also form covalent bonds with itself resulting in carbon chains.
Characteristics of organic compounds	Today organic compounds are generally considered to be compounds containing carbon. In contrast, inorganic compounds are those which contain elements other than carbon. The key difference between organic and inorganic compounds lies in bonding. Organic compounds are generally covalent and very stable. Inorganic compounds are generally ionic and often chemically reactive.
Related topics	The periodic table (A1) Nature of chemical bonding (B2)

Occurrence of carbon

All the elements of the periodic table are important for the balance between living and dead things on this planet to be maintained (see Topic G1). However, carbon is the most important of all. To illustrate, if all carbon containing compounds were removed, the Earth would be as barren as the moon appears. All that would remain is a small residue of minerals and water; as indeed found on the moon.

Carbon is a component of all food stuffs, fuels (including wood, gas, and oil), and fabrics (including clothing materials, paper, rubber, etc.).

At the mid-point of the twentieth century, over 6 million compounds containing carbon had been identified. This number is growing at the rate of around 1000 new compounds being synthesized or isolated from natural sources each year.

As a result of carbon being found in all living things, chemists up until 1828 thought that it was impossible to make organic (living) compounds in the laboratory, rather some **'vital force'** must be required. A significant turning point in this branch of chemistry came almost by accident in that year. Friedrich Wohler on heating a sample of an inorganic salt, ammonium cyanate, produced **urea**. As urea is a compound found in both blood and urine, it is unquestionably organic (see *Fig. 1*).

This preparation of an organic compound from inorganic reagents prompted others to attempt to synthesize further organic compounds, and to the rapid expansion in the number of known organic materials.

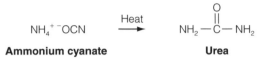

Ammonium cyanate   **Urea**

Fig. 1.   The synthesis of urea from ammonium cyanate.

Bonding characteristics of carbon

There are two major reasons for the huge number of organic compounds. The first relates to the bonding characteristics of carbon, the second to the isomerism of carbon containing compounds; dealt with in Topic E1.

Carbon is a member of Group IV of the **periodic table**, with a maximum **valency** of 4 (see Topics B1 and A1) and the ability to form regular **tetrahedral structures**. As the first member of the group carbon has an **electronegativity** (see *Fig. 2* and Topic A1), which is in the middle of the electronegativity scale. Therefore, it is generally considered as being neither electropositive nor electronegative, and hence is ideally suited for inclusion in **covalent** compounds (see Topic B2). Being in the middle of the electronegativity scale, carbon is able to form covalent bonds with atoms of higher (e.g. oxygen) and lower (e.g. hydrogen) electronegativity and of course with itself. Indeed the ability to form chains and to branch into the 3-dimensions of the tetrahedral structure is a very important feature of carbon chemistry.

H	2.3				
Li	0.9	Na	0.9		
Be	1.6	Mg	1.3		
B	2.1	Al	1.6		
C	2.5				
N	3.1	P	2.3		
O	3.6	S	2.6		
F	4.2	Cl	2.9	Br	2.7

Fig. 2.   Key elements of the periodic table and their electronegativity values.

Characteristics of organic compounds

In defining a compound as being organic or inorganic, it is generally accepted that the presence or absence of carbon is being indicated; organic compounds being carbon based. There are a number of exceptions to this definition. For example, carbon monoxide (CO), carbon dioxide (CO_2), carbonates, and cyanides, are all classified as inorganic as they were originally obtained from nonliving systems. Over and above this, organic and inorganic compounds differ in many of their physical and chemical characteristics (see *Table 1*) many of these resulting from the fact that organic molecules are generally held together by **covalent bonds** and inorganic compounds by **ionic bonds** (see Topic B2).

Table 1.   Properties typical of organic and inorganic compounds (in chemical rather than biochemical environments)

Property	Organic compounds	Inorganic compounds
Bonding **within** molecules	Usually covalent	Often ionic
Forces **between** molecules	Generally weak	Generally strong
Physical state	Gas, liquid or low melting point solid	Usually high melting point solid
Solubility in water	Generally low	Usually high
Rate of chemical reactions	Usually low	Usually high
Flammability	Often flammable	Usually nonflammable

D2 Naming organic compounds

Key Notes

Names of simple hydrocarbons	The simplest hydrocarbons (hydrogen and carbon containing compounds) are methane (CH_4), ethane (C_2H_6), propane (C_3H_8), and butane (C_4H_{10}), their names have an historical basis. Longer chain hydrocarbons are named more systematically, depending on the number of carbon atoms and using the Greek word for those numbers.
Naming branched hydrocarbons	The naming system for simple hydrocarbons is the basis for the naming of all organic compounds. When branches are introduced to a molecule their positions are referred to relative to the longest carbon chain, and the 'branched group' named in the same manner as a straight chain molecule.
Names of key functional groups	The main centers for reactivity in organic compounds are referred to as functional groups. Compounds containing specific functional groups are discussed together, as their chemistry is dictated by the nature of the functionality. In the context of biology, perhaps the most important functional groups are -OH (hydroxyl), -C=O (carbonyl), -P-O, P=O (phosphoro), and -C-N (amine). Groups can appear in combination and hence an extensive naming system is required.
Naming multifunctional molecules	When both branching and functional groups are present in a compound it is necessary to use a standardized naming system, that is the system devised by the International Union of Pure and Applied Chemistry (IUPAC). For all but the most complicated molecules, the IUPAC system permits unambiguous names to be applied. The basis for this system is the name of the longest (parent) hydrocarbon chain, and a prioritized list of functional group names, where the functional groups include hydrocarbon branches.
Related topics	Alcohols and related compounds (J1) Carboxylic acids and esters (J3) Aldehydes and ketones (J2) Amines and amides (J4)

Names of simple hydrocarbons

Straight chain hydrocarbons, more routinely referred to as **alkanes**, are for the most part named according to the number of carbons in the chain. The only exceptions to this are the four simplest alkanes, methane, ethane, propane and butane, whose names have historical roots. When there are more than four carbons in a straight chain then the molecule name is **prefixed** by the **Greek name for that number**, and **suffixed** by **-ane**, to indicate that the class of compound is alkane (see *Table 1*).

Naming branched hydrocarbons

The names of branched hydrocarbons or branched alkanes follows on logically from the names given for the straight chain **parent** molecule. To assign a name the longest carbon chain is selected and the carbons in this chain numbered, such that the positions of branches carry the lowest position number possible (see *Fig. 1*).

When more than one branch is present the same methodology is applied (see *Fig. 2*).

Table 1.   Names of straight chain alkanes

Number of carbons (n)	Name	Formula (C_nH_{2n+2})
1	Methane	CH_4
2	Ethane	C_2H_6
3	Propane	C_3H_8
4	Butane	C_4H_{10}
5	Pentane	C_5H_{12}
6	Hexane	C_6H_{14}
7	Heptane	C_7H_{16}
8	Octane	C_8H_{18}
9	Nonane	C_9H_{20}
10	Decane	$C_{10}H_{22}$
11	Undecane	$C_{11}H_{24}$

Names of key functional groups

The most reactive sites in organic molecules are termed **functional groups**. These are often sites in which carbon is bound to an atom other than another carbon or hydrogen; that is a heteroatom. However, carbon-carbon double and triple bonds are also referred to as functional groups as these are also susceptible to attack

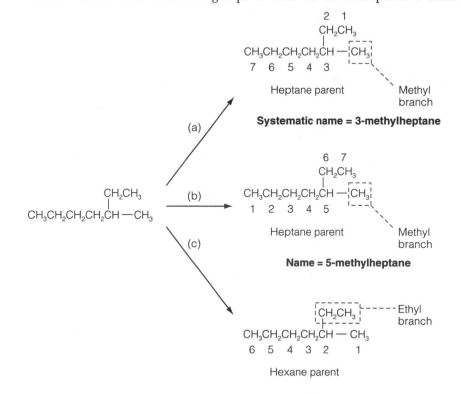

Fig. 1.   An illustration of how a systematic name may be applied. Only the name assigned in route (a) is correct.

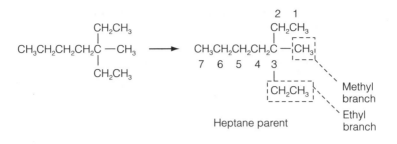

Systematic name = 3-ethyl-3-methylheptane

*Fig. 2. Systematic naming of hydrocarbons containing more than one branch point. Note the ethyl group takes priority over the methyl group in the name applied; in a similar way butyl takes priority over propyl, ethyl and methyl. The names for the branches, termed **alkyl** groups, stem from their parent alkane name.*

during the course of a chemical reaction. The nature of the heteroatom and the bond to the carbon atom dictates the reactivity of a compound, over and above (in general) the length of the alkyl chain and consequently compounds containing like functional groups need to be named to reflect this. In *Table 2* a list of the route structures, class name and name ending (to be used in conjunction with the names for hydrocarbons), is provided for some of the most important functional groups.

Table 2. Functional groups presented by class name, structure and name ending. When a bond connection is not specified it may be assumed that either a carbon or hydrogen atom may be attached

Class name	Route structure	Name ending
Alkene	$\mathrm{\diagdown C = C \diagup}$	- ene
Alkyne	$-C \equiv C-$	- yne
Alcohol	$-$ OH	- ol
Ether	$-$ OR	ether
Amine	$- NR_2^1$	- amine
Nitrile	$-$ CN	-nitrile
Sulphide	$-$ SR	sulphide
Thiol	$-$ SH	-thiol
Aldehyde	$-\overset{\overset{\displaystyle O}{\|\|}}{C}-H$	- al
Ketone	$-\overset{\overset{\displaystyle O}{\|\|}}{C}-R$	- one
Carboxylic acid	$-\overset{\overset{\displaystyle O}{\|\|}}{C}-OH$	- oic acid
Carboxylic acid Ester	$-\overset{\overset{\displaystyle O}{\|\|}}{C}-OR$	- oate
Amide	$-\overset{\overset{\displaystyle O}{\|\|}}{C}-NR_2^1$	-amide

R = alkyl, R^1 = alkyl or hydrogen.

Naming multifunctional molecules

When more than one type of branching group is present in a molecule, problems might arise in deciding how to name the compound. However, the International Union for Pure and Applied Chemistry (IUPAC) system largely takes care of this. The key point to remember is that the systematic name is made up of a **prefix** which specifies where each branch is and the name of the branching group (presented in the order shown below), the central part of the name stems from the **parent** alkane, and finally a **suffix** which is derived from the highest priority functional group.

As a general guide, the priorities of functional groups are as follows;

$$R < ^{\backslash}_{/}C=C^{/}_{\backslash} < -C\equiv C- < -OH < -\overset{\displaystyle O}{\overset{\displaystyle \|}{C}}-R < -\overset{\displaystyle O}{\overset{\displaystyle \|}{C}}-OH$$

R = alkyl or hydrogen

Some examples of multifunctional compounds and their systematic names are provided in Table 3.

Table 3. Some examples of the application of the IUPAC system for naming multifunctional organic compounds

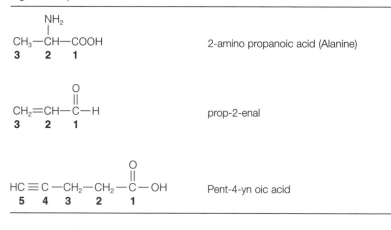

CH$_3$—CH—COOH with NH$_2$ on C2 (3 2 1)	2-amino propanoic acid (Alanine)
CH$_2$=CH—C—H with =O on C1 (3 2 1)	prop-2-enal
HC≡C—CH$_2$—CH$_2$—C—OH with =O on C1 (5 4 3 2 1)	Pent-4-yn oic acid

E1 ISOMERISM

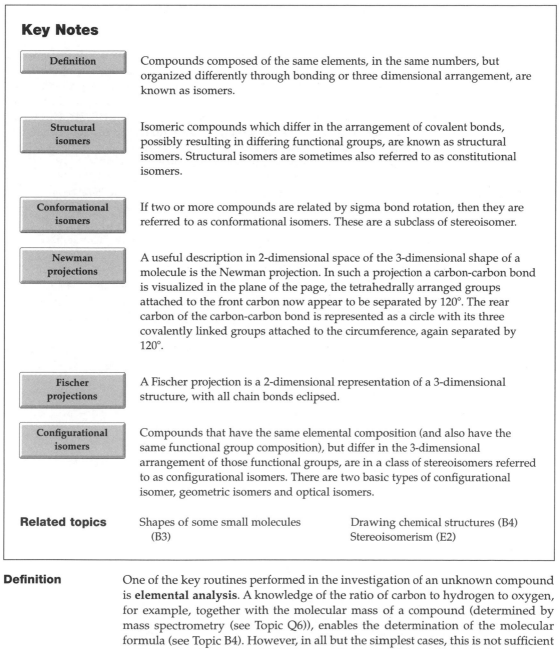

Key Notes

Definition
Compounds composed of the same elements, in the same numbers, but organized differently through bonding or three dimensional arrangement, are known as isomers.

Structural isomers
Isomeric compounds which differ in the arrangement of covalent bonds, possibly resulting in differing functional groups, are known as structural isomers. Structural isomers are sometimes also referred to as constitutional isomers.

Conformational isomers
If two or more compounds are related by sigma bond rotation, then they are referred to as conformational isomers. These are a subclass of stereoisomer.

Newman projections
A useful description in 2-dimensional space of the 3-dimensional shape of a molecule is the Newman projection. In such a projection a carbon-carbon bond is visualized in the plane of the page, the tetrahedrally arranged groups attached to the front carbon now appear to be separated by 120°. The rear carbon of the carbon-carbon bond is represented as a circle with its three covalently linked groups attached to the circumference, again separated by 120°.

Fischer projections
A Fischer projection is a 2-dimensional representation of a 3-dimensional structure, with all chain bonds eclipsed.

Configurational isomers
Compounds that have the same elemental composition (and also have the same functional group composition), but differ in the 3-dimensional arrangement of those functional groups, are in a class of stereoisomers referred to as configurational isomers. There are two basic types of configurational isomer, geometric isomers and optical isomers.

Related topics
Shapes of some small molecules (B3)

Drawing chemical structures (B4)
Stereoisomerism (E2)

Definition
One of the key routines performed in the investigation of an unknown compound is **elemental analysis**. A knowledge of the ratio of carbon to hydrogen to oxygen, for example, together with the molecular mass of a compound (determined by mass spectrometry (see Topic Q6)), enables the determination of the molecular formula (see Topic B4). However, in all but the simplest cases, this is not sufficient to determine the nature of the compound present. This is because, while obeying the rules of valence (see Topic A2), it is generally possible to write more than one structure for a particular molecular formula. For example, the molecular formula C_2H_6O is that of ethanol (CH_3CH_2OH) and dimethylether (CH_3OCH_3).

Although ethanol and dimethylether have the same elemental composition they are not the same compound, they are isomers (structural or constitutional isomers to be precise).

There are two basic classes of isomer; **structural** and **stereoisomer** (see Topic E2 for labeling of such isomers), conformational and configurational isomers are subclasses of the latter, and it is important that these are understood to identify an unknown unambiguously.

Structural isomers

The term **structural isomer** is the most general term that may be used to describe any compound which has the same molecular formula as another compound, but different physical properties. The example of ethanol and dimethylether has already been provided; these are in different chemical classes. Propan-1-ol ($CH_3CH_2CH_2OH$, boiling point 97.4°C) and propan-2-ol ($CH_3CH(OH)CH_3$, boiling point 82.4°C) are both C_3H_8O alcohols but are structural isomers. They can not be interconverted by bond rotation, and they are not related as mirror images. That is they do not fall into the other categories of isomer.

Conformational isomers

Carbon is a tetravalent atom capable of forming four covalent bonds. Once formed, these bonds must point towards the corners of a tetrahedron (see Topics B1 and B3). However, once more than one tetrahedral center is present, there are a number of ways in which these tetrahedrons may be organized with respect to each other. Take as a simple example ethane with its two sp^3 carbon centers. Several structures may be drawn which differ merely in degree of rotation about the carbon-carbon bond. Two extreme situations may be envisaged; one in which all of the bonds are in line with each other (i.e. **eclipsed**), the other in which all of the bonds are as far away from each other as possible (i.e. **staggered**) (see *Fig. 1*).

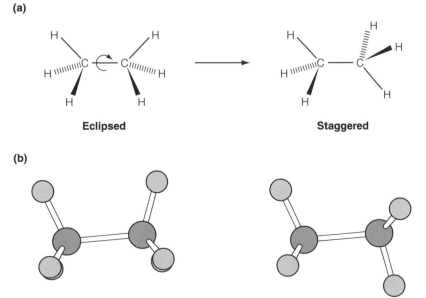

(a)

Eclipsed   Staggered

(b)

Fig. 1.   The (a) eclipsed and (b) staggered structures of ethane.

These are the two extreme '**conformations**' of ethane, they differ in energy by approximately 2.8 kcal mol^{-1}. In the eclipsed conformation, the proximity of the hydrogen atoms on the two carbon centers leads to unfavorable electronic and steric repulsions, making this a high energy conformation. In contrast, in the staggered conformation these interactions are reduced and this is therefore an energetically favorable arrangement.

As two conformations differ in the 3-dimensional arrangement of atoms, conformational isomers are also stereoisomers.

Newman projections

When illustrating the 3-dimensional shapes of molecules, the diagrams as used in *Fig. 1* for ethane can become very cumbersome. A much more straightforward method for conveying the same information is however available. In the 1930s M.S. Newman developed a convention for representing molecules in 2-dimensional space. In drawing the so-called **Newman projection** for the staggered conformation of ethane, the carbon-carbon bond is placed in the plane of the page. The front carbon is represented as a dot with three arms (the carbon-hydrogen bonds), separated in the plane of the page by 120°. The rear carbon is represented by a circle with three arms (C-H bonds) attached to its circumference, separated by 120°. The angle between the C-H bonds on the front carbon and those on the rear (the dihedral angle) is 60° (see *Fig. 2*).

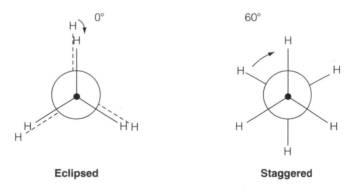

Eclipsed   **Staggered**

Fig. 2.   Newman projections of the eclipsed and staggered conformations of ethane.

This convention is readily applied to larger molecules, as illustrated in *Fig. 3* for the conformational isomers of butane.

Clearly the addition of two 'methyl' groups to the ethane framework leads to an increase in the number of possible conformations and new labels must be introduced. Now the most energetically favored conformation is that with the methyl groups as far apart as possible. To distinguish this from any other staggered arrangement it is labeled *anti*. The eclipsed conformation in which the methyl groups are in line with each other is labeled *syn*, this is the highest energy conformation. The range of energies encompassed in conformational variations is readily depicted in **an energy profile** (see *Fig. 4*).

Conformational isomerism is also possible in cyclic systems. For example, cyclohexane (C_6H_{12}), can adopt a range of conformations *via* bond rotations which occur to relieve the **strain energy** (or torsional strain) that would be present if the ring was forced to be planar.

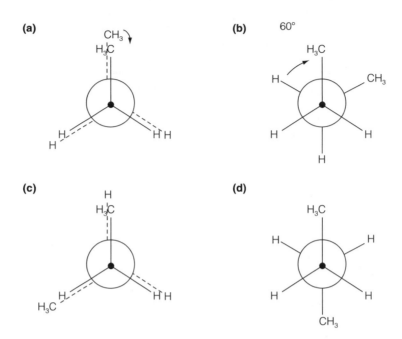

Fig. 3.   The Newman projections for the conformational isomers of n-butane. (a) syn-eclipsed, (b) gauche, (c) eclipsed, (d) anti.

The energy profile, with a representation of the distinct conformational isomers of cyclohexane is shown in *Fig. 5(a)*, together with the Newman projections of the (i) chair and (ii) boat conformations of cyclohexane (see *Fig. 5(b)*).

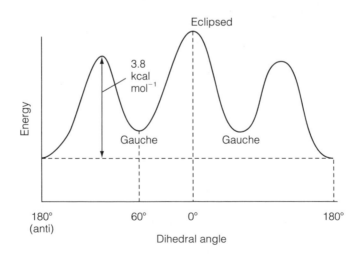

Fig. 4.   The energy profile for conformational flexing in n-butane.

Fischer projections

The Fischer projection is another way to display, in two dimensions, the 3-dimensional shape of a molecule. For example the Fischer projection for butane is shown in *Fig. 6(a)*.

(a)

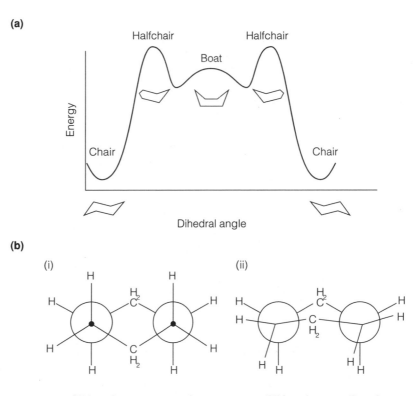

(b)

(i)   (ii)

All bonds are staggered   **All bonds are eclipsed**

Fig. 5.   (a) The energy profile for conformational flexing in cyclohexane, and (b) the Newman projections for the (i) chair and (ii) boat conformations.

(a)

$$CH_3$$

H———H

H———H

$$CH_3$$

Eclipsed

(b)

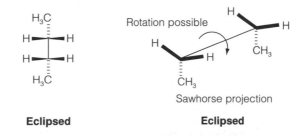

H_3C

H━━━━H

H━━━━H

H_3C

Rotation possible

Sawhorse projection

Eclipsed   **Eclipsed**

Fig. 6.   (a) The Fischer projection for n-butane and (b) the 3-dimensional shape it represents.

Configurational isomers

The horizontal and vertical lines are reflecting the **eclipsed** conformation by convention. The horizontal lines represent bonds which in the 3-dimensional structure would come out of the plane of the paper, the vertical lines represent bonds that would be behind the plane of the paper (see *Fig. 6(b)*).

The 3-dimensional shape of the chair and boat conformations of cyclohexane are clearly very different, they are stereoisomers of cyclohexane. However, they are interconvertable *via* a series of bond rotations. Stereoisomers that cannot be interconverted by bond rotations but rather by bond breaking and bond forming, are referred to as **configurational isomers**. For example, there are two **stereoisomers** of but-2-ene (see *Fig. 7*).

Fig. 7.   Two stereoisomers of but-2-ene. (a) Boiling point 3.7°C. (b) Boiling point 0.9°C.

As rotation about a carbon-carbon double bond is not possible (under normal conditions) (7a) and (7b) may only be interconverted by the breaking and making of two covalent bonds. Compounds (7a) and (7b) belong to the subclass of configurational isomers referred to as **geometric** isomers (see Topic E2). Such stereoisomers have different physical properties and if present as a mixture may readily be separated. Another example of configurational isomerism is offered by butan-2-ol (see *Fig. 8*). Carbon atom number 2 has four different groups attached to it and is referred to therefore as an **asymmetric** center. This carbon is also referred to as a **chiral** center as there are two ways in which the four groups (H,

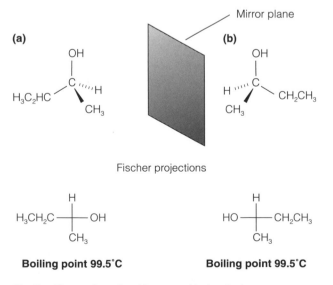

Fig. 8.   The configurational isomers of butan-2-ol.

OH, CH_3, CH_2CH_3), may be arranged about the asymmetric carbon and these two ways are related as nonsuperimposable mirror images.

As (8a) and (8b) are mirror images their structures are nonsuperimposable and it is only possible to convert from (8a) to (8b) by breaking and making two covalent bonds. Unlike geometric configurational isomers, these chiral compounds do not have different physical properties and therefore are not readily isolated from intimate mixtures. The only way in which (8a) and (8b) differ, is in the direction in which they rotate plane polarized light (see Topics E2 and E3). Consequently (8a) and (8b) are often referred to as **optical isomers**.

E2 STEREOISOMERISM

Key Notes

Geometric isomers	Stereoisomeric compounds which have different physical properties and cannot under normal conditions be interconverted *via* bond rotations, are referred to as geometric isomers.
E and Z isomers	The labels E and Z are applied to geometric isomers that are tri- or tetrasubstituted, that is when the *cis/trans* labeling system breaks down. To assign these labels a priority, which is determined by atomic number, has to be attributed to each substituent. The Z label (equivalent to *cis*) means the highest priority substituents are on the same face of the molecule, the E label (equivalent to *trans*) means they are on opposite faces.
Enantiomers	An asymmetric, or chiral, carbon can adopt two structures. These structures are mirror images of each other and are referred to as enantiomers. These have identical physical and chemical properties, they rotate plane polarized light by equal amounts but in opposite directions.
R and S isomers	The R and S labels are attached to chiral centers and unambiguously describe the 3-dimensional arrangement of substituents around the chiral center. To assign these labels, substituents have to be prioritized by virtue of atomic number.
Diastereoisomers	Compounds containing more than one chiral center may exist in pairs of enantiomers or diastereoisomers. Diastereoisomers are not mirror images of each other, and have different physical properties. Mixtures of diastereoisomers can be separated.
D and L isomers	The D and L labels are given to chiral compounds, traditionally to naturally occuring biochemically important compounds. They are not unambiguous and they do not indicate the direction in which light is rotated. The labels are applied relative to a standard compound.
Related topics	Isomerism (E1) Optical activity and resolution (E3)

Geometric isomers

Compounds which only differ in 3-dimensional structure by virtue of the relative positions of sets of substituents on a carbon-carbon double bond, or carbon-carbon bond constrained in a ring, are referred to as geometric isomers. An example of geometric isomerism is offered by but-2-ene (see Topic E1). The isomer with the methyl groups on the same side of the double bond is labeled *cis*, that with the methyls on opposite faces is labeled *trans* (see *Fig. 1*). Note that although other structures can be drawn for the same molecular formula (e.g. 2-methylpropene), these are not geometric isomers of but-2-ene.

Geometric isomers involving ring systems are labeled in the same way. There are two possible structures which may be draw for 1,2-dimethylcyclohexane (see

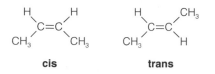

cis   **trans**

Fig. 1. Geometric isomers of but-2-ene.

Fig. 2). Under normal conditions of temperature and pressure, rotation about a carbon-carbon double bond is inhibited and it is only possible to convert one geometric isomer to another by a series of bond breaking and bond remaking reactions. Similarly, although conformational changes can take place in cyclic systems, it is only possible to change the relative arrangements of substituents by the breaking of covalent bonds. Geometric isomers therefore have different physical and chemical properties and are isolatable.

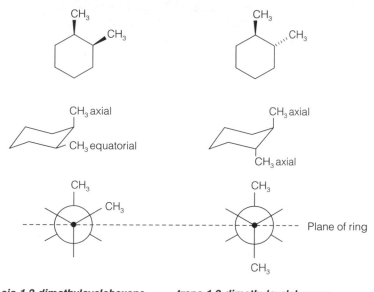

cis-1,2-dimethylcyclohexane   *trans-1,2-dimethylcyclohexane*

Fig. 2.   Three different representations of the geometric isomers of 1,2-dimethylcyclohexane.

E and Z isomers   When geometric isomers are tri- and tetrasubstituted (hydrogen is not considered as a substituent) the *cis/trans* labels are no longer adequate. An alternative labeling system is used which differentiates between and prioritizes substitutents on the basis of atomic number, such that substituents of high atomic number have high priority. Thus, the two geometric isomers of 1-bromo, 1-chloro, 2-fluoroethene are as shown in *Fig. 3*.

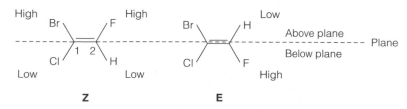

Fig. 3.   Assigning E and Z labels to 1-bromo, 1-chloro, 2-fluoroethene.

To specify a particular isomer it is necessary to prioritize the substituents, Cl, F, Br, and determine the relative positions.

	Br	>	Cl	>	F	>	H
Atomic number	35		17		9		1

High priority   Low priority

At carbon 1, in the left hand structure of *Fig. 3*, Br has a higher atomic number than Cl, Br therefore has higher priority. At carbon 2, F has priority over H. Therefore, in the left hand structure the two highest priority substituents are on the same plane and this molecule is labeled **Z**, from zussamen, the German for together (like *cis*).

Conversely, in the right hand structure the two highest priority groups are on opposite planes, therefore this molecule is labeled **E**, from entgegen, German for opposite. If the atomic numbers of the first attached atoms are the same then simply move to the next bond along (see *Fig. 4*).

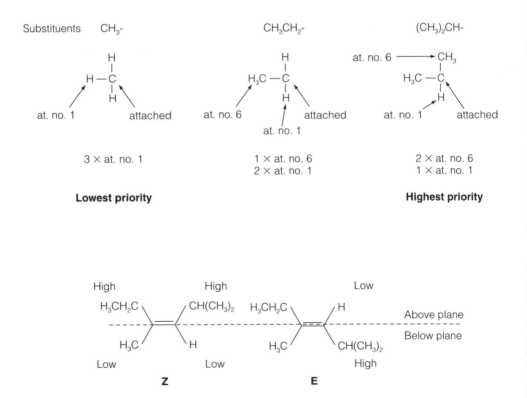

Fig. 4. Further examples of priority assignment for E and Z determination.

Hence, the *E* and *Z* isomers of 2,4-dimethylhex-3-ene are as shown in *Fig. 4*. If the attached group is multiply bonded to other atoms, treat each bond as a separate substituent (see *Fig. 5*).

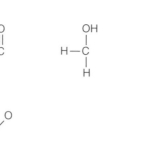

2 × at. no. 8   1 × at. no. 8
1 × at. no. 1   2 × at. no. 1

Higher priority   **Lower priority**

Low priority –H < –CH$_3$, –CH$_2$CH$_3$ < –CH = CH$_2$ < –Ph < CH$_2$OH < –CHO < –COR <
–COOH < –COOR < –NH$_2$ < –NR$_2$ < –OH < –F   **High priority**

Fig. 5. Priority assignment of groups containing multiple bonds, and other common functional groups Ph, benzene ring; R, alkyl group.

Enantiomers

A pair of compounds each with one or more chiral centers which are mirror images of each other is known as an **enantiomer**. Such compounds are identical in all ways except that they rotate plane polarized light in opposite directions. For example, lactic acid (a) is the enantiomeric partner of lactic acid (b) (see *Fig. 6*). Lactic acid (a) is found (see Topic E3) to rotate plane polarized light in a clockwise direction. It would therefore be labeled as the (+), or (*d*) enantiomer; '*d*' from dextrorotatory meaning **clockwise**. Lactic acid (b) rotates light anticlockwise and would therefore be labeled (−) or (*l*), from levorotatory meaning **anticlockwise**.

A **1 : 1 mixture** of *d*- and *l*-lactic acid would produce no **net** rotation of plane polarized light. A mixture of enantiomers in this ratio is called a **racemic mix** (see Topic E3). Many natural products and drugs have chiral centers and can therefore exist as enantiomeric pairs, but frequently only one enantiomer is recognized by the natural product, enzyme, or drug target, due to the specificity of 3-dimensional structure in biological interactions.

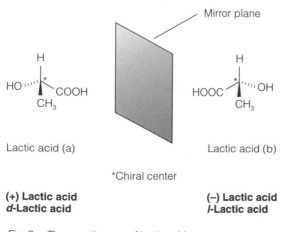

Lactic acid (a)   Lactic acid (b)

*Chiral center

(+) Lactic acid   **(−) Lactic acid**
***d*-Lactic acid**   ***l*-Lactic acid**

Fig. 6.   The enantiomers of lactic acid.

Lactic acid (a)   HO>HOOC>H$_3$C>H
                                    1   2   3   4

1.

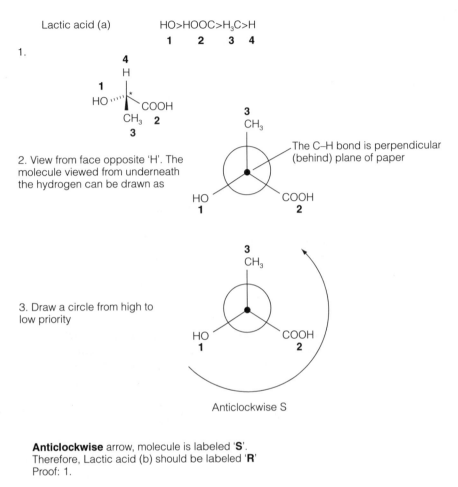

2. View from face opposite 'H'. The molecule viewed from underneath the hydrogen can be drawn as

The C–H bond is perpendicular (behind) plane of paper

3. Draw a circle from high to low priority

Anticlockwise S

Anticlockwise arrow, molecule is labeled '**S**'.
Therefore, Lactic acid (b) should be labeled '**R**'
Proof: 1.

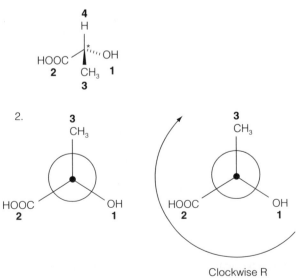

Clockwise R

Fig. 7.   Assignment of R and S labels to the enantiomers of lactic acid.

R and S isomers

R and S are labels given to chiral centers, as distinct from *d* and *l*. Knowing that a particular center is S, enables the drawing of a single 3-dimensional shape; it is an unambiguous labeling system. An S form of a molecule has the R form as its enantiomeric partner. We know that the R and S isomer of a particular compound will rotate light in opposite directions, however, we cannot predict which direction they will rotate; experiments are required for this (see Topic E3). Thus, in general it is possible to have R(+) and R(−) isomers.

The labels R and S are applied following implementation of the **Cahn-Ingold-Prelog rules**:

● Assign priority to the groups around the chiral center, label these from 1 to 4, 1 being the highest.
● View the molecule from the face opposite the group of lowest priority, draw a circle through groups 1, 2, and 3.
● If the direction of drawing the circle is **clockwise**, then label the chiral center **R** (rectus) , if **anticlockwise** label the center **S** (sinister).
● Repeat for each chiral center.

In *Fig. 7* these rules are applied to the enantiomers of lactic acid.

The labels R and S (and D and L, d (+) and l (−)) are labels of **configuration**. In the case of lactic acid the **R-configuration** rotates light by **−3.82°**, i.e. anticlockwise; therefore this isomer may also be labeled (−) or **l**, and the **S-configuration** rotates light by **+3.82°**, i.e. clockwise; this isomer may also be labeled (+) or **d**.

Diastereoisomers

When two or more chiral centers are present in a molecule it is possible to draw pairs of enantiomers or diastereoisomers. These have identical compositions but are not mirror image related, and indeed have different physical properties (see *Fig. 8*). Each compound in *Fig. 8* has identical substituents;

(8a) and (8b) at C-1 NH_2, COOH, $CH(OH)CH_3$, H
(8a) and (8b) at C-2   OH, $CH(COOH)NH_2$, CH_3, H

However, (8a) and (8b) are not mirror images of each other, and indeed as they would therefore have different physical properties they could be separated from each other in a mixture.

To prove that (8a) and (8b) are not mirror images the configurations, R and S, of the chiral centers need to be determined. Following the procedure described in *Fig. 7*, C-1 in (8a) has the 'R' configuration as does C-2; this compound would be labeled 1R,2R, and its enantiomer would be labeled 1S,2S. In compound (8b) C-1 is

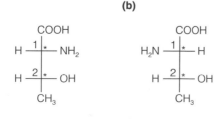

(a)

(b)

*=Chiral carbon

Fig. 8.   *Fischer projections of two molecules related as diastereoisomers. (a) D-allothreonine* $[\alpha]_D^{26} - 9.6°$. *(b) L-threonine* $[\alpha]_D^{26} - 28.3°$.

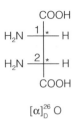

$[\alpha]_D^{26}$ 0

*Fig. 9.   An example of a **meso** compound. The optical rotation due to chiral center (1) is equal and opposite to that due to chiral center (2), the net optical rotation is therefore zero; an internal racemic mix.*

in the 'S' configuration, but C-2 is in the 'R' configuration, this compound is therefore labeled 1S,2R, and is clearly not the mirror image of compound (8a).

Diastereoisomers are optical isomers and will rotate plane polarized light. Under special circumstances, however, it is possible that the net optical rotation will be zero; an **intramolecular racemic mix** (see *Fig. 9*)

In *Fig. 9* carbon-1 and C-2 have the same substituents, therefore they will rotate light by the same magnitude. The configuration of C-1 is 'S' and of C-2 'R', therefore chiral center 1 will rotate light in the opposite direction to chiral center 2. The net result is zero rotation of light. Compounds of this type are referred to as **meso** forms.

D and L isomers   The labels D and L are rarely used today, except by biologists when referring to natural amino acids and sugars (see Topic K1). They are assignments of configuration relative to a standard compound; the standard usually being glyceraldehyde (see *Fig. 10*).

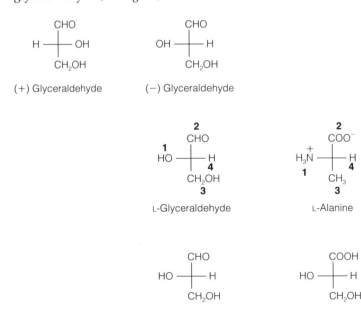

*Fig. 10.   D and L labels of 'relative' configuration. (The numbers refer to the priorities of the different functional groups, **1** being the highest, **4** the lowest).*

(+)-Glyceraldehyde is designated the 'D' configuration, and (−)-glyceraldehyde the 'L' configuration. Subsequently structures which had similar layouts to the (+) isomer were labeled 'D', those similar to the (−), were labeled L. An example from the common amino acids is shown in *Fig. 10* to illustrate this point.

The numbers 1 to 4 indicate priority from high to low. Note that the arrangement of priorities is the same for L-alanine as it is for L-glyceraldehyde. This system however breaks down, as it does not take account of the chemical nature of the substituents. Consequently it is possible to have a 'D' configuration which proves to be levorotatory (−) as shown for the oxidation product of L-(−)-glyceraldehyde, the L-(+)-carboxylic acid, 2,3-dihydroxypropanoic acid (see *Fig. 10*).

E3 OPTICAL ACTIVITY AND RESOLUTION

Key Notes

Polarimetry	Polarimetry is used to determine the effect on plane polarized light of enantiomeric compounds.
Specific rotation	Specific rotation is the degree of rotation of plane polarized light taking account of sample concentration and light path length.
Optical purity	A sample of a single enantiomer is said to be optically pure.
Resolution	The separation of enantiomers from mixtures is referred to as resolution.

Related topics Isomerism (E1) Circular dichroism (Q5)
  Stereoisomers (E2)

Polarimetry

Enantiomeric compounds (see Topics E1 and E2) have identical physical properties, except that they rotate the path of plane polarized light in opposite directions. The device used to measure the degree of rotation of light is called a **polarimeter** (see *Fig. 1*). It consists of a light source in which light is oscillating in all possible directions. This is passed through a 'polarizer' or polaroid filter, and the light emerging travels in one direction only. This polarized light then passes through the polarimeter tube containing a solution of the enantiomer. As light passes through the solution it is rotated. At the end of the tube is an analyzer which detects the extent to which the light has been rotated (the optical rotation).

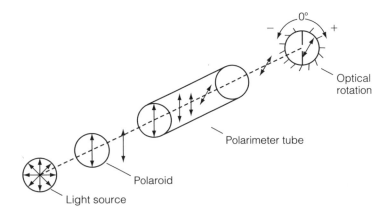

Fig. 1. Schematic diagram of a polarimeter.

Specific rotation   The degree to which plane polarized light is rotated by an optically active compound is referred to as the **optical rotation**, and is given the symbol α. However, in addition to the actual nature of the enantiomeric compound there are several factors which can influence α. The wavelength of the original light source, the temperature, sample concentration, type of solvent, together with the actual length of the polarimeter tube all effect the magnitude of the optical rotation. Consequently, rather than quoting 'optical rotation' values it is more usual to quote the **specific rotation**, $[\alpha]$. By definition the specific rotation is the rotation detected using a 10 cm tube, and sample concentration of 1 g ml^{-1};

$$[\alpha] = \frac{\alpha}{c \cdot l}$$

where **c** is the concentration in **grams per milliliter** and l the path length in **centimeters**.
The nature of the light source is usually indicated as follows;

$$[\alpha]_D^{20°}$$

where **D** refers to the **sodium D line** and 20° is the **temperature** in celsius units.

Optical purity   A sample of a single enantiomer is said to be optically pure. Nature, for the most part, uses and generates optically pure compounds, but chemists rarely do. The most common result of a chemical reaction that produces a chiral center is a mixture of enantiomers. If the mixture is **1 : 1** in enantiomers, a **racemic** mixture is said to be the result (see Topics E1 and E2), and no net optical rotation will be detected. However, it is usually the case that an optical rotation is detected, but not of the size expected; that is, found previously by experiment for the isolated enantiomers. This is due to one of the enantiomers being in excess. The **enantiomeric excess** (ee) is given by;

$$\text{enantiomeric excess (ee)} = \left(\frac{\text{moles of one enantiomer} - \text{moles of other enantiomer}}{\text{total number of moles}}\right) \times 100$$

The **ee** is often quoted in the synthesis of chiral compounds. However, it can only be determined when the enantiomers have been separated.

Resolution   The separation of enantiomers is described as **resolution**, the general idea of which was pioneered by Louis Pasteur in the mid-1800s.
There are three basic ways in which enantiomeric mixtures may be resolved;

(a) By mechanical means if the products are crystalline and not present in a 1 : 1 ratio.
(b) Enzymatically, one of the enantiomers will be recognized and converted and the other will remain in solution.
(c) By the production of diastereoisomers, as these have different physical properties. Traditional separation procedures may then be applied.

Pasteur originally separated crystalline enantiomers of sodium ammonium tartrate (see *Fig. 2*), by simply observing that two crystal forms were present and using tweezers to isolate each form. This is not a method in general use today.
   Some classes of enantiomeric compounds can act as substrates for enzymes. As in general only one stereoisomer would be recognized and turned over (or

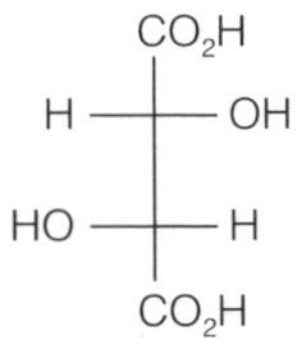

(+)-Tartaric acid

Fig. 2.   The Fischer projection of (+)-tartaric acid.

metabolized) this is a way of obtaining a single pure enantiomer. For example, as shown by Pasteur, *Penicillium galucum* (a green mold) metabolizes the (+)-enantiomer of sodium ammonium tartrate, leaving the (−)-enantiomer untouched and available for isolation. The preparation of diastereoisomers is perhaps the most common approach to resolution. A mixture of enantiomers is reacted with the single enantiomer of another reagent, the result is diastereoisomers. See *Fig. 3* for an illustration of this approach with acidic chiral compounds.

(+)-enantiomer

            +   (−)-base ⟶   (+)-enantiomer(−)base   (−)-enantiomer(−)base

(−)-enantiomer

 1 mole each   2 moles   1 mole   1 mole

                                               Diastereoisomers

Fig. 3.   The conversion of enantiomers to diastereoisomers is an important approach to resolution.

As diastereoisomers have different physical properties they can be separated using standard analytical techniques. Once separated they can of course be converted back to the desired enantiomer by removal of the base group.

F1 PHOSPHORIC ACID AND PHOSPHATES

Key Notes

Chemical properties	Phosphoric acid is a deliquescent solid, generally encountered as a viscous aqueous solution. It is weakly acidic, with three possible sequential deprotonation steps, forming phosphates. Like carboxylic acids, phosphoric acid can dimerise *via* a dehydration reaction to form phosphoanhydrides.
Occurrence in biology	In biological systems phosphoric acid is usually present in its ionized form, that is as a phosphate. Anhydrides of phosphoric acid are common, occuring in adenosine triphosphate (ATP) and adenosine diphosphate (ADP). Since ATP releases phosphate to form ADP and liberate energy, phosphate appears in many metabolic processes. Phosphate also occurs in the 'building blocks of life', the nucleic acids, as a phosphodiester.
Related topics	Oxygen (F3) Oligonucleotide synthesis (M2) Carboxylic acids and esters (J3)

Chemical properties

Phosphoric acid (H_3PO_4) is often referred to as **orthophosphoric acid**. Like its ionized equivalents, the phosphates or **orthophosphates** ($H_2PO_4^-$, HPO_4^{2-}, PO_4^{3-}), phosphoric acid is a tetrahedral compound (see *Fig. 1*). As a result, just like tetrahedrally substituted carbon, the phosphorus of the acid has the potential for chirality (see Section E). Phosphoric acid itself may be obtained as a semi-solid, but is more usually available as a viscous aqueous solution. The viscosity arises from the ability of phosphoric acid to hydrogen bond; having three hydrogen bond donation sites and one hydrogen bond acceptance site (see *Fig. 1* and Topic H1).

Phosphoric acid is referred to as being **tribasic**, in that it has three possible dissociation steps (see Topic N3). As a result, mono- and disodium and/or potassium salts of phosphoric acid (see Topic N1) are routinely used as pH buffers for *in vitro* biochemical studies and indeed play an important part in the buffering of biological processes *in vivo*.

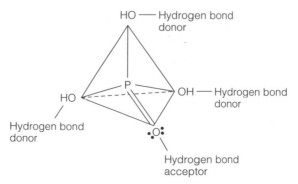

Fig. 1. Phosphoric acid; tetrahedral structure and hydrogen bonding sites.

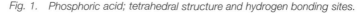

One of the most important reactions of phosphoric acid and its derivatives, in terms of biology, is that of multimerization. As with carboxylic acids (see Topic J3) two phosphoric acid molecules may combine, with the loss of water, to form a diphosphate ester, also referred to as **pyrophosphate**. However, as phosphoric acid has further 'OH' functionalities, triphosphates may also be formed.

Salts of phosphoric acid are solid and many are relatively water insoluble unless a strong mineral acid is present. Salts are routinely formed between phosphate and sodium, but more importantly in terms of biology, calcium and magnesium phosphates (see Topic G3) are also common.

Occurrence in biology

Phosphate ions are present, in the form of calcium phosphate, in both bones and teeth, lending to each their rigidity. Phosphate ions are found in every cell and are the primary anions of the intercellular fluid. To illustrate their importance consider the breakdown of **glycogen**. Glycogen is essentially a polymer of glucose (see *Fig. 2*). At times when glucose is in short supply, immediately before meals or during exercise, glycogen is metabolized to meet the glucose demand (glucose being a key energy source for the body). The pathway for glycogen breakdown was first elucidated by Carl and Gerty Cori and was found to involve a phosphorylation step utilizing orthophosphate (which may also be referred to as **inorganic phosphate**, P_i) (see *Fig. 3*). The reaction is of course enzyme catalyzed, by **glycogen phosphorylase**, and results in the formation of a phospho-sugar; glucose-1-phosphate. This process is energetically favorable due to the formation of the phosphorylated sugar which, due to its charge, cannot diffuse out of the cell (see Topic H2). In contrast a hydrolytic cleavage of the glycosidic bond (see *Fig. 3*) would yield glucose which is able to cross the cell membrane.

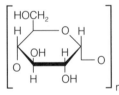

Fig. 2.   Glycogen, a glucose polymer.

Phosphate ions and phosphate esters are extremely important in the synthesis and breakdown, respectively, of adenosine triphosphate (ATP) (see *Fig. 4*). ATP is considered as the 'universal currency of free energy' (see Topic O4) in biological systems. It is continuously formed and consumed. Consequently phosphate ester bonds are continuously being formed and broken and phosphate ions continuously generated and consumed (see *Fig. 4*).

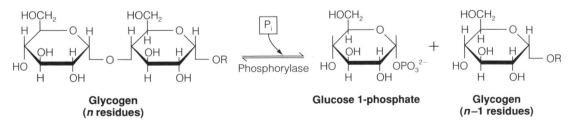

Fig. 3.   Glycogen metabolism involves a phosphorylation step.

(a)

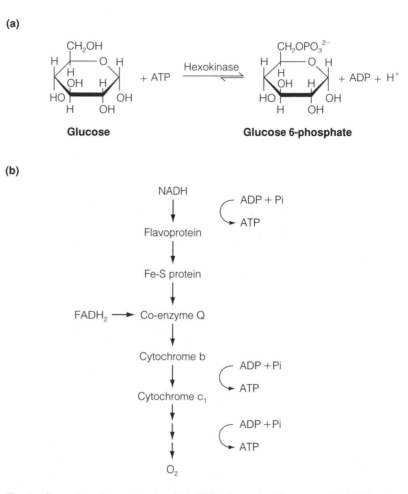

(b)

Fig. 4.   Examples of processes in which ATP is formed and consumed. (a) Utilization of ATP in the phosphorylation of glucose. (b) Formation by oxidative phosphorylation (see Section G).

Another important site where phosphate groups (more precisely phosphodiesters) may be found is in the nucleic acids (see Topic M2), where they provide the link between the ribose rings of successive nucleosides.

F2 NITROGEN

Key Notes

The N_2 molecule

Nitrogen is the major constituent of the Earth's atmosphere. The nitrogen molecule consists of two atoms (so it is sometimes called dinitrogen), which are joined by a formal triple bond. This is an especially strong bond, which makes the molecule relatively stable and chemically inert.

Biological importance

Nitrogen is an important component of many biological molecules. The conversion of atmospheric nitrogen to other compounds is termed *fixation*. Despite its chemical inertness, N_2 fixation is performed by specialist enzymes in certain bacteria. Nitrogen is returned to the atmosphere by the action of denitrifying microorganisms in soil.

Oxides of nitrogen

N_2 in the atmosphere is converted to various nitrogen oxides by the action of lightning and solar radiation. Such compounds are also generated by a number of industrial and combustion processes, and are important components of urban and industrial air pollution. Nitrogen dioxide is an especially dangerous pollutant molecule, due to its toxicity and reaction with water to form nitric acid. Oxides of nitrogen are important components of photochemical smog. Nitrous oxide is also known as laughing gas and has anesthetic properties. Nitric oxide is an important signaling molecule for a wide range of physiological processes.

Solubility in water

Nitrogen is slightly soluble in water under normal conditions, but the solubility rises as the applied pressure is increased. This means increased nitrogen is dissolved in blood and body tissues when normal air is breathed under high pressure, as can occur in deep sea diving. This is the underlying cause of 'the bends' and nitrogen narcosis. The latter is related to the tendency of nitrogen to dissolve in the fatty components of nerve and brain cells.

Related topics

Nature of chemical bonding (B2)   First law of thermodynamics
Magnesium and manganese (G3)   (O2)

The N_2 molecule

The major constituent of the Earth's atmosphere is nitrogen, in the form of the N_2 molecule. This is sometimes called 'dinitrogen', but usually just 'nitrogen' or 'the nitrogen molecule'. By volume, nitrogen makes up about 78% of the atmosphere. It is a particularly stable, chemically inert molecule. We can represent the nitrogen molecule as follows:

$$N\equiv N$$

that is, with a triple bond between the two atoms. The stability of the molecule is related to the relatively high strength of this triple bond. The bond energy of the $N\equiv N$ molecule (the energy required for homolytic cleavage to produce two nitrogen atoms, see Topic B3) is 942 kJ mol^{-1}, which compares to 432 for the H-H bond in the H_2 molecule.

We can also describe the bonding between the atoms in terms of molecular orbitals. We have previously used this approach to analyze the bonding in carbon monoxide (section B2). Indeed, the situation here is closely analogous to that of carbon monoxide, because the same number of electrons is involved: the CO and N_2 molecules are said to be **iso-electronic**. We can construct a set of molecular orbitals just as we have previously. So, combinations of the 2s atomic orbitals form sigma bonding (σ) and anti-bonding ($\sigma*$) molecular orbitals. A 2p atomic orbital from each atom also combine to form σ and σ^* molecular orbitals, whilst the other 2p atomic orbitals (2 in each atom) form π and π^* orbitals; this is shown in *Fig. 1*.

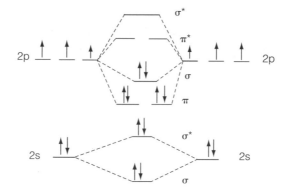

Fig 1.   MO diagram for N_2

By allocating the electrons to the molecular orbitals by the aufbau principle (see Topic A2), we find that all the p-electrons are accommodated in bonding orbitals (σ and π), whilst none (of the 'p' electrons) are in antibonding orbitals (σ^* and π^*). The bonding is thus optimal; we have two strongly connected atoms.

Biological importance

Nitrogen is of enormous biological importance. Nitrogen atoms are constituents of many biological molecules, such as proteins and nucleic acids (see Topics K1, K2). Although the atmosphere is rich in nitrogen, under normal conditions it is not a particularly reactive molecule, and most organisms cannot satisfy their need for nitrogen directly from the atmosphere.

Fortunately, a biological process for accessing atmospheric nitrogen has evolved. The reacting of atmospheric nitrogen to form other compounds is called **fixation**. Nitrogen from the atmosphere is converted to ammonia, as well as nitrites and nitrates, by the action of specialist enzymes. One such enzyme contains a molybdenum atom at its center (section G4). Such enzymes are expressed in nitrogen-fixing bacteria. There are two classes of these: the free-living variety, and those living in symbiotic relationships with plants. These bacteria are found in root nodules of many plants, where plant and bacteria exist in intimate association. The plant provides an appropriate environment for the bacteria, which in turn supplied nitrates and other nitrogenous compounds for the plant's metabolic needs. Although the action of lightning and sunlight also consume atmospheric nitrogen (see below), bacteria are responsible for over 90% of total global nitrogen fixation.

Nitrogen fixation would eventually mean the depletion of the atmospheric reserve of nitrogen, but this is prevented by the action of **denitrifying micro-**

Oxides of nitrogen

organisms, which live in soil. These convert nitrates to N_2, which finds its way back into the atmosphere. This is an important link in the **nitrogen cycle**.

Under the action of lightning discharges, and of sunlight, atmospheric nitrogen reacts with another atmospheric component, the reactive oxygen molecule O_2 (see Topic F3), to form a range of compounds collectively termed nitrogen oxides. These can also be formed when combustion is occurring at particularly high temperatures, which allows N_2 to react with O_2. The generation of oxides of nitrogen is thus often associated with industrial activity and fuel combustion.

The various oxides of nitrogen generally have individual names for general usage, but may also be referred to by the oxidation state (see Topic I2) of nitrogen, which by convention is written in Roman numerals. Some oxides of nitrogen are listed in *Table 1*; it should be noted that a number of other nitrogen oxides exist, but they generally have little biological importance.

Table 1. Some oxides of nitrogen

Name	Formula	Alternative name	Comments
Nitrous oxide	N_2O	nitrogen(I) oxide	'laughing gas'
Nitric oxide	NO	nitrogen(II) oxide	messenger molecule
Nitrogen dioxide	NO_2	nitrogen(IV) oxide	brown color
Nitric acid	H_2NO_3		strong acid

Nitrous oxide is also known as laughing gas, and is used as an anesthetic. It can induce a mild hysteria, and is also an option for pain relief during labor.

Nitric oxide is a colorless, toxic gas and is unusual in that it is a stable compound with an odd number of electrons; it is a radical. Although traditionally regarded as a poison, in recent years nitric oxide has unexpectedly been found to be an important signaling molecule, acting in a wide range of tissues and physiological processes. Within tissues, the molecule is relatively short-lived (a few seconds), and is generated by specialist enzymes called nitric oxide synthases. Because of the large number of processes in which NO seems to be involved, it may well be that a number of diseases involve faulty regulation of NO synthesis.

Nitrogen dioxide is a pungent and irritating gas, and is one of the more dangerous components of industrial and urban atmospheric pollution. It causes the build-up of fluid in the lungs (pulmonary edema). It reacts with water at low temperatures thus:

$$2NO_2 + H_2O \rightarrow HNO_3 + HNO_2$$

to form a mixture of nitric (HNO_3) and nitrous (HNO_2) acid. This is an example of **disproportionation**: a compound giving rise to two products, one in a higher oxidation state, and the other in a lower oxidation state, than the original compound (see Topic I2). Nitrous acid is unstable, and readily decomposes (again, disproportionates) to form nitric oxide and nitric acid:

$$3HNO_2 \rightarrow HNO_3 + 2NO + H_2O$$

Thus, at higher temperatures, nitrogen dioxide gives rise to nitric acid and nitric oxide. The formation of nitric acid is of special concern, since when this occurs in the atmosphere, it contributes to acid rain.

Nitric acid is a strong acid (see Topic N2). It forms salts, nitrates, which are important agricultural fertilizers. These can be added to soil, but as they are generally highly water soluble they can be carried away in run-off water, causing environmental problems in draining rivers and the sea.

Oxides of nitrogen are important components of photochemical smog, which can form over urban areas. Mixtures of hydrocarbons and oxides of nitrogen undergo a number of reactions which are catalyzed by sunlight. Amongst the products are highly toxic ozone (see Topic F3), and NO_2, which gives the distinctive light-brown haze of this serious form of pollution. If they find their way into the upper atmosphere, nitrogen oxides (especially nitrous oxide) may cause depletion of the ozone layer (see Topic F3). The nitrogen oxides are fragmented by the action of ultra-violet radiation to produce atoms and free radicals (see Topic I1), which then react with ozone molecules.

Solubility in water

The solubility of dinitrogen in water decreases as the temperature is increased; this is also the case for O_2, and is why bubbles of air can often be seen forming and emerging from water as it is heated. This reduced solubility with rising temperature arises because the dissolution of N_2 (and of O_2) is an example of an exothermic process (see Topic O2), a characteristic of which is that it becomes less favorable with increased temperature.

The solubility of N_2 in water is not particularly high under normal conditions. At 25 °C and 1 atmosphere pressure of air, the solubility of nitrogen is around 6.8×10^{-4} mol/l. This is about 18 mg of nitrogen, which would occupy about 15 cm^3 as a gas. As the N_2 molecule has no dipole moment it interacts with other molecules via what can be described as hydrophobic interactions (see Topic H2). Consequently, nitrogen is more soluble in fatty tissue and lipids (see Topic K3) than it is in water.

Although the solubility of nitrogen in water is not high, it is a very important factor in undersea diving. This provides a good illustration of the interaction of a gas with human tissue under varying pressures. The water pressure quickly rises as a diver descends from the surface. This pressure is exerted on the diver and the gas that is being breathed. If the diver is breathing normal air (i.e. mostly nitrogen), increasing amounts of nitrogen will dissolve in the diver's blood (and passed on to other body tissues, especially fatty tissue). Problems will then arise if the return to the surface, and lower pressure, is too rapid. This causes nitrogen to come out of solution and form bubbles in the blood; the painful and dreaded 'bends'. This can be avoided if the return to the surface is very slow, or a process of slow decompression is carried out. This allows dissolved nitrogen to gradually be lost from the body without forming bubbles. An alternative is to breathe mixtures of oxygen and helium. Helium is chemically inert (see Topic A1), and less soluble in water than nitrogen. Consequently, under pressure less helium will dissolve in the blood and tissues than would nitrogen, so the process of decompression can be quicker. A related situation can arise in aviators in un-pressurized aircraft, who may suffer from the bends at high altitude.

Another danger of breathing normal air at high pressure is nitrogen narcosis, a state of impaired mental function, which can be highly dangerous and disturbing. We have seen above that nitrogen dissolves well in fatty tissue. Nerve and brain cells contain high levels of lipids (Topic K3), and the presence of increased nitrogen dissolved in these components causes the condition.

The problems caused by nitrogen solubility are not confined to deep sea diving using breathing apparatus. A condition known as taranava syndrome has been observed in Japanese and Polynesian pearl divers, who repeatedly dive to around 50 m, whilst holding their breath. This repetition mean that nitrogen builds up in tissues; the syndrome is avoided if sufficient time is allowed for the nitrogen to escape from the body between dives.

F3 OXYGEN

Key Notes

The O_2 molecule	A molecular orbital analysis of the O_2 molecule shows that it has two electrons in π anti-bonding orbitals. The lowest energy arrangement of these electrons has them in separate orbitals with their spins aligned. This makes the molecule a di-radical, which causes O_2 to be highly reactive.
Biological importance of oxygen	The reactivity of O_2 is such that at chemical equilibrium, there would be little if any present in the Earth's atmosphere. However, O_2 is constantly being generated by photosynthesis. Oxygen is an absolute requirement for much of life on Earth. It is necessary for the oxidation of carbohydrates to produce high-energy ATP, which is then used in general metabolic processes. Oxygen is transported from the lungs by becoming bound to hemoglobin and myoglobin within red blood cells. Away from the lungs, the gaseous pressure of O_2 is low and the oxygen affinity of hemoglobin and myoglobin markedly decreases, allowing oxygen to be released to the tissues.
Solubility of oxygen in water	Oxygen is soluble in water to an extent that supports oxygen-breathing aquatic life. The solubility of oxygen in water decreases markedly with increasing temperature, because the dissolution of oxygen is an exothermic process. Oxygen is highly soluble in fluorocarbons, emulsions of which can be used as synthetic blood.
Ozone	Under the action of sunlight or electrical discharge, O_2 gives rise to ozone, O_3, which is another allotrope of oxygen. Ozone is highly toxic and dangerous if generated at ground level, but in the upper atmosphere it is an important absorber of solar ultra-violet radiation, providing important protection for ground-dwelling life. Ozone can react with certain chlorofluorocarbons due to photochemical reactions. These compounds can diffuse to the upper atmosphere and so may deplete the ozone there.
Related topics	Molecular orbitals (B1) Buffers (N4) Reactive species (I1)

The O_2 molecule

The bonding of the two atoms of the oxygen molecule can be analyzed using the molecular orbital approach, as we have seen for CO (see Topic B2) and N_2 (see Topic F2). A diagram of the energy levels of the molecular orbitals in O_2 is given in *Fig. 1*. It resembles those for CO and N_2, except that spectroscopic analysis and detailed calculations show that the relative energies of the σ and π molecular orbitals formed by the 2p atomic orbitals are different in N_2 and O_2.

Allocating electrons to the molecular orbitals, according to the aufbau principle (see Topic A2), we find that once all the lower orbitals have been filled, we are left with two electrons. By the standard rules, these should be placed in π* orbitals. There are two of these, which are of equal energy (i.e. degenerate). These two orbitals arise because side-on overlap is possible between two pairs of atomic

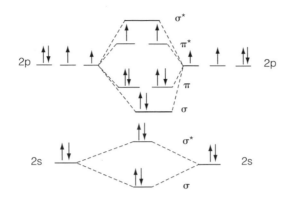

Fig. 1. Molecular orbital representation for O_2

p-orbitals. The lowest energy distribution of these two electrons is to place one in each of these π^* orbitals, with their spins parallel. This is due to Hund's rule (see Topic A2). This treatment predicts that the O_2 molecule behaves as a di-radical (i.e. two unpaired electrons, see Topic I1). The molecule should also be **paramagnetic**, which is a form of magnetism which is due to the presence of unpaired electrons in a molecule. These predictions turn out to be correct. The two unpaired electrons mean that the molecule is highly reactive.

The O_2 molecule can be written as O=O, i.e. with a double bond between the two atoms, but this is not a very good description. Another representation is $O^\bullet{=}O^\bullet$, where the dots indicate the presence of the unpaired electrons.

Biological importance of oxygen

Oxygen is crucial to life on Earth. Humans can only survive for a matter of minutes without breathing.

Around 21% of the Earth's atmosphere consists of oxygen, as the O_2 molecule. However, oxygen is highly reactive, and if conditions of chemical equilibrium were to prevail, there would be little if any oxygen in the atmosphere. This is not the case because oxygen is continually generated biologically, by photosynthesis (see Topic G3). The presence of significant O_2 in the atmosphere means that ozone is generated (see below), which is also of great importance to life.

The requirement for oxygen arises principally from the metabolic need for the oxidation of carbohydrates (sugars) (see Topic F1) to produce useable energy. The overall reaction is:

$$C_6H_{12}O_6 \ + \ 6O_2 \ + \ 36ADP + 36P_i \ \rightarrow \ 6CO_2 \ + \ 36ATP + 6H_2O$$
glucose

This process is the sum of a series of reactions, and molecules other than glucose can also be oxidized.

The energy is stored in the ATP molecule (see Topic F1). The hydrolysis of ATP to yield P_i (inorganic phosphate) and ADP (adenosine-5'-diphosphate) or AMP (adenosine-5'-monophosphate) is highly favorable in terms of Gibbs free energy (see Topic O4), which is then available for metabolic purposes. In general metabolism, the hydrolysis of ATP is often coupled to other reactions that would be thermodynamically unfavorable in isolation.

In higher organisms, O_2 is transported to and from organs and tissues by specialist oxygen-binding proteins, hemoglobin and myoglobin (see Topic G2), located within red blood cells. Although oxygen is, to some extent, soluble in water (see below), the general demand for oxygen is such that it cannot be met from the oxygen dissolved in the aqueous component of blood. A litre of blood can absorb about 250 cm^3 of oxygen by virtue of its hemoglobin, but only about 3 cm^3 of oxygen can be dissolved in its aqueous component.

The ability of hemoglobin and myoglobin to transport oxygen is related to their affinities for the O_2 molecule, and how these are related to the pressure of gaseous oxygen. This is shown for hemoglobin in *Fig. 2*.

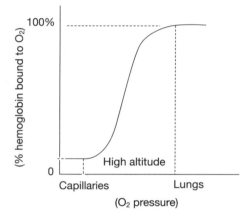

Fig. 2.   Simple plot of hemoglobin/O$_2$ affinity.

At high O_2 pressure, as in the lungs, hemoglobin is effectively **saturated** by oxygen, i.e. 100% (or nearly so) of the hemoglobin is bound to O_2. At lower O_2 pressure, as in the blood capillaries, the affinity is reduced and so O_2 is released from the hemoglobin, and is able to diffuse into tissues. At altitude, the pressure of O_2 is reduced, and as indicated in *Fig. 2*, hemoglobin leaving the lungs may not be saturated. This means less oxygen will be available to body tissues, which may become starved of oxygen; altitude sickness can result. Given time, the body will respond to this situation by increasing the number of red blood cells. But the hemoglobin will remain only partially saturated under these conditions; evidently it has evolved at low altitude.

Solubility of oxygen in water

Aquatic creatures are able to breathe oxygen because it is soluble in water. The solubility of O_2 in water is about twice that of N_2 (see Topic F2). As with N_2, the solubility of O_2 decreases as the temperature increases, because the dissolution of the gas is an **exothermic** process. Such a process becomes less favorable at higher temperature (see Topic O2). This means that the amount of dissolved oxygen in water nearly halves when the temperature is raised from 5 to 35 °C. Consequently, in hot weather fresh water fish can face difficulties.

The solubility of oxygen in water increases when the gas pressure is raised. This is also the case with nitrogen (see Topic F2), which causes problems in deep sea diving. Under high pressure, an increased amount of oxygen *dissolves* in the blood. If the pressure is rapidly relieved, in principle we would expect the formation of oxygen bubbles within tissues and blood cells, just as happens with nitrogen. However, this does not occur with oxygen; the diver's metabolic needs, which are increased by the exertion of the dive, seem adequate to consume any extra oxygen present in the tissues. This is fortunate, because unlike nitrogen, oxygen is a component of the air that cannot be replaced by any other gas!

It has been found that oxygen is soluble in fluorocarbons (hydrocarbons in which the hydrogen atoms have been replaced by fluorine). The solubility is such that animals immersed in these fluids can, if persuaded, breathe normally. These fluorocarbons are not water-soluble, but can be made into emulsions in water (see Topic C1), and used as blood substitutes during surgery. This synthetic blood can be useful when the patient has a rare blood type not supported by blood banks.

Ozone

Ozone is a form of oxygen which is triatomic, i.e. there are three atoms in the molecule. The atoms are arranged such that the molecule resembles the letter 'V'. O_3 and O_2 are allotropes of oxygen. This means that they are different forms of the same, pure, element. Another example is graphite and diamond, which are allotropes of carbon.

Ozone is formed from O_2 by the action of ultraviolet radiation, or by electrical discharge. Ozone has a distinctive smell, which can be detected after thunderstorms and around electrical equipment. It is a pale blue gas, and is actually highly toxic. It can also be generated by certain forms of air pollution. The action of sunlight on nitrogen oxides mixed with hydrocarbons (as in photochemical smog, see Topic F2) produces ozone, which being near ground level constitutes a severe health risk. However, ozone is an important component of the upper atmosphere, and is crucial to life on Earth as it now exists.

Under the action of ultraviolet light (wavelengths below 242 nm), O_2 molecules are dissociated into atoms:

$$O_2 \rightarrow O + O$$

When it encounters another O_2 molecule, an oxygen atom combines with it to form ozone:

$$O + O_2 \rightarrow O_3$$

The ozone molecule is readily decomposed by ultraviolet radiation of wavelengths shorter than 300 nm. The net result of this photochemical creation and destruction of ozone is that there is a layer of the upper atmosphere that is relatively rich in ozone. This is called the **ozonosphere**, or more commonly 'the ozone layer', and is the region between the altitudes of 10 and 50 km. The presence of the ozonosphere effectively blocks solar radiation of wavelength shorter than 290 nm, so that it does not reach the ground. In the absence of this shielding by ozone, this radiation would be deadly.

In recent years there has been considerable concern about the potential damage to the ozone layer by chlorofluorocarbons (often termed CFCs). These are chlorine- and fluorine-containing organic compounds, developed mainly for use as refrigerants. These compounds can diffuse up to the ozone layer, where they encounter ultraviolet radiation. This causes the molecules to decompose, for example:

$$CCl_2F_2 \rightarrow CClF_2 + Cl$$

The chlorine atom is highly reactive (more so than the $CClF_2$ radical), and reacts with ozone thus:

$$Cl + O_3 \rightarrow ClO + O_2$$

In this way, ozone can be depleted. Any damage to the ozone layer would have profound consequences, and so the use of CFCs is now seriously discouraged.

Alternative refrigerants include the hydrofluorocarbons (HFCs), for example 1,1,1,2-tetrafluoroethane. These compounds are more reactive than are CFCs, which means that they are destroyed before they are able to rise up to the altitude of the ozone layer.

Nitrogen oxides (see Topic F2) have also been implicated in ozone depletion. It seems that for the most part these compounds do not reach the upper atmosphere in significant quantities, except possibly nitrous oxide. However, this would not be the case for nitrogen oxides generated *in situ* by stratosphere-cruising supersonic aircraft. This has generated a contentious debate, but for the time being there are very few such vehicles.

G1 THE EARLY TRANSITION METALS

Key Notes

Chemical and physical properties	Key features of the early transition elements include their high melting points and favorable mechanical properties; making these industrially important metals. Other features, of greater significance in biology, include their ability to display variable oxidation states and catalytic activity.
Relative abundance of transition metals	Many metals are essential for biology; amongst the transition metals these include iron, cobalt, zinc, copper, vanadium, chromium, manganese, nickel, and molybdenum. Of these only molybdenum is in relatively short supply. Life on this planet has evolved to use the most abundant elements.
Porphyrin complexes	Heavy metals, like the transition metals, are generally maintained in biological systems complexed with organic ligands. Perhaps the most important of the organometallic complexes in this context is that involving a metal ion and a porphyrin ring. A porphyrin is a tetrapyrrole system, constructed to retain the aromaticity of pyrrole itself. Two of the four nitrogens of the pyrroles are covalently linked to a metal divalent cation, the other two nitrogens coordinate with the metal *via* their electron lone pairs. Porphyrins can accommodate a range of ions, the precise properties of the ion-porphyrin complex being dictated in part by the nature of substituents attached to the tetrapyrrole system.
Related topics	The periodic table (A1) Magnesium and manganese (G3) Nature of chemical bonding (B2) Cobalt and molybdenum (G4) Iron (G2)

Chemical and physical properties

The original definition of a transition metal was an element or ion with a partially filled d-orbital (see Topic A1). This definition is somewhat out of date, as today discussions of the transition elements include, for example, zinc, copper, silver and cadmium all of which have complete outer d-orbitals. Elements now classified as transition elements are thus grouped by virtue of their chemical and physical characteristics.

The transition elements are extremely important in industrial processes, valued for their high melting points and mechanical strength. The metals are in general very dense. This may be accounted for by their relatively small atomic radius; scandium (the largest atom in the first transition series) has a radius of 1.44 Å compared with 1.74 Å for chlorine. There is little variation in atomic radius across a transition series.

The majority of transition metal ions are colored (see *Table 1*); the exceptions, in the first series, being scandium and zinc having an empty and full d-shell respectively and hence no electrons which can be excited (see Topics Q1 and Q2). A feature of transition metal chemistry is their ability to form coordination compounds (see Topic B2). The nature of coordinated ligands determines the color detected for

Table 1. Outer electronic configurations of some transition metals; oxidation states and ion colors

Outer electronic configuration	Oxidation states (stable state in bold)	Ion color (in aqueous solution)
Sc $3p^6\ 3d^1\ 4s^2$	**3**	Sc^{3+} colorless
Ti $3p^6\ 3d^2\ 4s^2$	**4** 3 2	Ti^{3+} purple
V $3p^6\ 3d^3\ 4s^2$	5 **4** 3 2	V^{3+} green
Cr $3p^6\ 3d^5\ 4s^1$	6 **3** 2	Cr^{3+} violet
Mn $3p^6\ 3d^5\ 4s^2$	7 6 4 **3** **2**	Mn^{3+} violet, Mn^{2+} pink
Fe $3p^6\ 3d^6\ 4s^2$	6 **3** 2	Fe^{3+} yellow, Fe^{2+} green
Co $3p^6\ 3d^7\ 4s^2$	4 3 **2**	Co^{2+} pink
Ni $3p^6\ 3d^8\ 4s^2$	4 **2**	Ni^{2+} green
Cu $3p^6\ 3d^{10}\ 4s^1$	**2** 1	Cu^{2+} blue
Zn $3p^6\ 3d^{10}\ 4s^2$	**2**	Zn^{2+} colorless

a particular complexed ion. The majority of transition metal ion complexes contain six ligands surrounding the central ion in an **octahedral** arrangement. However, other complexes occur which have a **tetrahedral** or **square planar** arrangement (see *Fig. 1*). These complexes involve a combination of covalent bonds and 'electrostatic interactions' or **dative** bonds (see Topic B2).

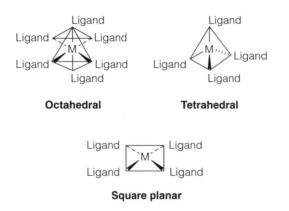

Octahedral   **Tetrahedral**

Square planar

Fig. 1. Shapes of coordination compounds formed by transition elements (M, metal).

Key to the biological importance of the transition metals is the ability to display **variable oxidation states** (see *Table 1*). A range of oxidation states is available as there is only a gradual increase in the ionization energies for successive ionizations; for nontransition metals there is a noticeable gap in the energy required for successive ionization steps;

Transition metal

$$V \xrightarrow{648} V^+ \xrightarrow{1,364} V^{2+} \xrightarrow{2,858} V^{3+} \xrightarrow{4,634} V^{4+}$$

Nontransition metal

$$Al \xrightarrow{578} Al^+ \xrightarrow{1,811} Al^{2+} \xrightarrow{2,745} Al^{3+} \xrightarrow{11,540} Al^{4+}$$

Energy in kJ mol^{-1}

In general the lower oxidation states are reducing and the higher oxidation states are oxidizing. Coupled with variable oxidation states is the ability of transition metals to catalyze chemical reactions; catalysis often involves changing the oxidation state of the metal. For example iron (III) catalyzes the reaction between iodide and persulfate ions;

$$2Fe^{3+} + 2I^- \longrightarrow 2Fe^{2+} + I_2$$

$$2Fe^{2+} + S_2O_8^{2-} \longrightarrow 2Fe^{3+} + 2SO_4^{2-}$$

Similar processes occur when metals appear in biological systems (see Topics G2, G3 and G4).

Relative abundance of transition metals

Although there are over 100 elements in the periodic table only a relatively small number of these are of significance in biology; among these, elements may be classified as **essential** or **nonessential**. The definition of an essential element is straightforward; these elements are vital for life. The definition of a nonessential element is more vague and varies amongst species. Nonessential elements are often considered to be **trace** elements, that is, in short supply. However, one of the important elements in mammalian biology is molybdenum, a trace essential element!

Of the transition elements, those essential in biology include iron, cobalt, zinc, copper, vanadium, chromium, manganese, nickel, and molybdenum. Apart from molybdenum (already referred to) these are amongst the most abundant elements on the planet; iron is the fourth most abundant element (see *Table 2*). Indeed there are only four elements which are abundant but are of little relevance to biology; silicon, aluminium, titanium, and zirconium. It would therefore seem reasonable to suggest that life has evolved to utilize those elements most readily available. Support for this suggestion comes from consideration of organisms which have developed to handle the stress of unusually high or unusually low elemental abundance; as in hydrothermal vents where organisms survive in an environment of extremely high iron, copper and zinc concentrations dissolved from surrounding rocks by superheated water.

Porphyrin complexes

Many of the transition metals of importance in biology are stored as complexes with organic ligands. Amongst the most important examples of these so-called organometallic complexes are those which involve a porphyrin ring and a metal ion.

Table 2. Relative abundance of some elements of importance in biology

Element	Earth's crust (g kg^{-1})
Oxygen	474
Sodium	24
Magnesium	20
Phosphorus	1
Sulfur	0.26
Calcium	42
Manganese	0.95
Iron	56
Cobalt	0.025
Molybdenum	0.0015
Iodine	0.0005

A porphyrin ring is a macrocyclic system comprised of four pyrrole rings (see *Fig. 2*) connected one to the other *via* a carbon atom. A series of alternate or conjugated double bonds are present (see Topic L1) ensuring that the **aromatic** nature of pyrrole itself is maintained in the macrocycle. The nitrogens of the four pyrroles are involved in complexing the metal. Porphyrins can accept two hydrogens to form the diacid (see *Fig. 2*) or donate two protons to form the dianion. It is the latter that is the most important in biology, as the metal ions of interest are generally doubly positively charged.

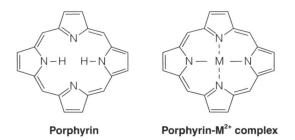

Porphyrin   **Porphyrin-M^{2+} complex**

Fig. 2.   The general porphyrin ring structure showing metal coordination site.

In complexing M^{2+} (M is a transition metal) two covalent bonds are formed and two 'coordinate' or dative bonds (see Topic B2) essentially hold the metal in the plane of the porphyrin ring. The hole in the centre of the porphyrin ring is such that all of the first row transition metals can be readily accommodated. Nickel is rather small, hence some puckering of the otherwise planar macrocycle occurs. In contrast, when larger metals are present these reside above the plane of the ring. As many of the transition metals are able to form octahedral complexes, a further two ligands may be present above and below the plane of the porphyrin ring (see Topics G2–G4).

The stability of porphyrin-M^{2+} complexes has been determined to follow the order ($Ni^{2+} > Cu^{2+} > Co^{2+} > Fe^{2+} > Zn^{2+}$). The kinetics of formation of the metalloporphyrins follows the order ($Cu^{2+} > Co^{2+} > Fe^{2+} > Ni^{2+}$). This is interesting as it suggests that cobalt porphyrins should be more prevalent than iron porphyrins. The fact that the opposite is true presumably depends on the fact that iron is over a thousand fold more abundant than cobalt.

Although the porphyrins that occur in biology have the same basic structure they differ in the nature of substituents attached to the pyrrole rings. It is these differences which enable fine tuning for particular metals and processes of importance in biology (see Topics G2–G4).

G2 IRON

Key Notes

Chemical and biochemical properties

Iron displays three oxidation states, the most stable of which being +3. It is most commonly found in the form of oxides and hydroxides. Iron hydroxides are relatively insoluble in water. To enable the uptake of iron by plants it is sometimes necessary to mask the iron in an organic complex. In the body, iron is stored as iron hydroxide particles surrounded by a proteinaceous coat. In higher animals iron is transported through the blood stream *via* transferrins. In microorganisms siderophores perform this task.

Heme proteins

Heme proteins contain one or more iron-porphyrin complexes. Many heme proteins perform vital roles in biology. Amongst such proteins are the cytochromes, critical in photosynthesis and respiration, and hemoglobin and myoglobin, required for oxygen storage and transport in blood.

Iron-sulfur proteins

Iron-sulfur proteins are those where iron is complexed *via* the sulfydryl group of cysteines and inorganic sulfurs or sulfides. Among the most important of iron-sulfur proteins are the ferredoxins and the related rubredoxins. These proteins are present in all green plants, algae and photosynthetic bacteria. Ferredoxins play an important role in nitrogen fixation.

Related topics

Oxygen (F3)
The early transition metals (G1)

Solubility (N5)

Chemical and biochemical properties

Pure iron is silver in appearance and melts at 1528°C. Iron will readily combine with oxygen, the halogens, nitrogen, sulfur and carbon on heating; all but the halogen compounds being of biological significance. Iron displays three oxidation states, the most stable being +3 (referred to as iron (III)); the +2 state is reducing (referred to as iron (II)) and the rare +6 state extremely oxidizing (see Topics G1 and I2). A simple chemical test for the presence of iron (III) ions in aqueous solution involves the use of potassium hexacyanoferrate (II); a dark blue precipitate known as **Prussian blue** forms;

$$K^+ \; + \; [Fe^{II}(CN)_6]^{4-} \; + \; Fe^{3+} \longrightarrow K^I \, Fe^{III}[Fe^{II}(CN)_6]^{4-}$$

Hexacyanoferrate(II)   Prussian blue

Iron is most commonly found in the form of oxides and hydroxides, the latter being very poorly water soluble. Indeed under conditions of low soil pH (see Topic N3) a number of plants have to obtain the iron they require for photosynthesis from synthetic organo-iron complexes. Nevertheless in the body iron is essentially stored as particles of iron (III) hydroxide known as **ferritin**; these are surrounded by a protein coat and maintained until required in, for example, **redox**

stream by proteins called **transferrins**. In microorganisms, iron solubilization and transport is achieved by low molecular weight proteins called **siderophores**.

Heme proteins

A number of extremely important proteins contain one or more **heme** groups, that is an iron-porphyrin (see Topic G1) complex, at their core. Examples of heme proteins include the **cytochromes**, vital for photosynthesis and respiration, and **hemoglobin** and **myoglobin** proteins required for the transportation of oxygen through the bloodstream and storage of oxygen in muscle, respectively.

One of the best understood cytochromes is cytochrome *c*. In this particular system the heme group has a polypeptide chain attached and wrapped around it. The iron is held in the porphyrin by interactions with the pyrrole nitrogens (see Topic G1), the fifth and sixth coordination sites being occupied by a histidine (of the polypeptide chain) nitrogen and the sulfur of a methionine residue (see Topic M1) (see *Fig. 1*). As all six coordination sites are occupied, cytochrome *c* cannot react *via* simple coordination but must react indirectly *via* an **electron-transfer** mechanism.

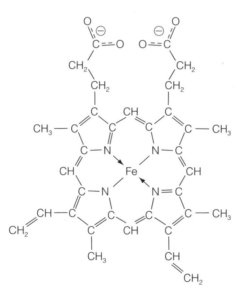

Fig. 1. Schematic representation of the arrangement of the heme group in cytochrome c.

There are a variety of cytochromes which display a range of **redox potentials** (see Topic I2), meeting the specific needs of a particular electron transfer scheme such as those in photosynthesis or respiration. These are cytochromes *a*, *b*, and *c*. The potentials are such that electron flow is from *b* to *c* to *a* to release molecular oxygen. Some of the '*a*' type cytochromes are thought to be capable of binding oxygen and reducing it; the central iron must therefore have only five of its six coordination sites occupied initially. This feature can account for the toxicity of cyanide ions (CN^-). Cyanide ions bind very tightly to iron and are not readily reduced, blocking this particular redox chain.

The prime function of hemoglobin is to transport oxygen from its source to the site of use within muscle cells. The oxygen is subsequently transferred to myoglobin for use in respiration.

The iron in both hemoglobin and myoglobin is in the +2 oxidation state. The iron in these proteins is in a similar environment to that of cytochrome *c*; the iron is coordinated to the nitrogen of a histidine in the protein chain. However, the pyrrole rings of the porphyrin are not linked to the protein, and only five of the six coordination sites are occupied; leaving room for oxygen to bind. Besides molecular oxygen, water, and carbon monoxide, can all act as ligands; carbon monoxide binding 50 times more tightly than oxygen to myoglobin and 200 times more tightly to hemoglobin. It is for this reason that prolonged carbon monoxide exposure can prove fatal.

Before the binding of the sixth ligand in either hemoglobin or myoglobin, the coordinated iron has been found to lie slightly above the plane of the porphyrin ring (see *Fig. 2*). On oxygen binding, for example, the iron drops into the plane of

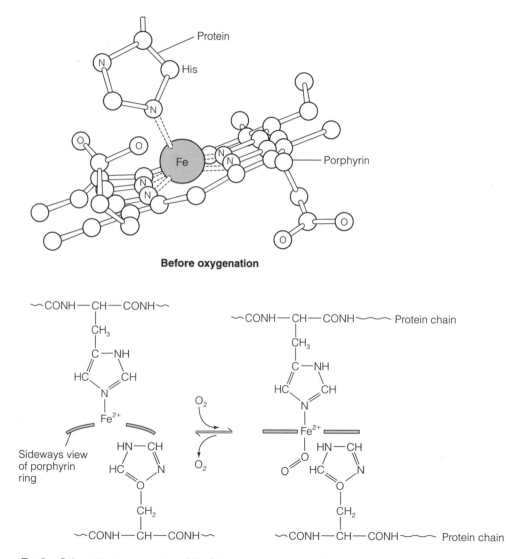

Fig. 2. Schematic representation of the heme group of hemoglobin and myoglobin prior to and following oxygen complexation.

the tetrapyrrole ring, pulling with it the protein chain (see *Fig. 2*). The protein regains its original structure when the oxygen molecule is released.

It should be noted that hemoglobin is a tetrameric system, that is, comprised of four globin polypeptide chains each chain having its own heme group. However, experiments have shown that these heme groups are not independent, but indeed cooperate on oxygen binding and release.

Iron-sulfur proteins

Many important proteins involve iron complexed in some way other than as a metalloporphyrin, amongst these the iron-sulfur proteins have been well studied. Rubredoxin and ferredoxin are iron-sulfur proteins found in plants and photosynthetic bacteria, and involved in photosynthetic transport as well as being involved in nitrogen fixation (see Topic G4).

The iron-sulfur clusters in these proteins involve iron ligated to the thiol group of cysteine residues, and other sulfur atoms or sulfides (in the case of ferredoxins). In each case the iron is tetrahedrally substituted, in some ferredoxins however these groups are linked together to form a cube arrangement (see *Fig. 3*).

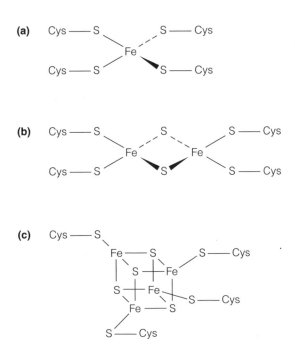

Fig. 3.   Iron-sulfur clusters of ferredoxin and rubredoxin. (a) From bacterial rubredoxin; (b) from photosynthetic ferredoxin; (c) from cubane like ferredoxin (s, inorganic sulfur).

The reduction of nitrogen in biology, or **nitrogen fixation**, to form ammonia is achieved by a complex of two enzymes; a **reductase** and a **nitrogenase**. Each of these involve an iron-sulfur protein, the latter also requires a molybdenum-iron complex (see Topic G4). Ferredoxin is the iron-sulfur protein in question (see *Fig. 3*).

G3 MAGNESIUM AND MANGANESE

Key Notes

Chemical properties

Magnesium as a Group 2A element displays an oxidation state of +2. Its compounds are generally ionic, although some display covalent character. Magnesium is commonly found as a carbonate or sulfate in minerals. Magnesium hydroxide is a strong base. Manganese is one of the first row transition metals and like other transition metals shows variable oxidation states, the +2 state being the most stable. Manganese is generally found in the form of an oxide. Manganese hydroxides are water soluble.

Magnesium in photosynthesis

One of the key components of photosynthesis is chlorophyll, a magnesium-porphyrin complex. There are two (main) chlorophylls, *a* and *b* differing only in the nature of one substituent on the porphyrin ring. The function of chlorophyll is to absorb light, making this light available as the energy source for a cascade of oxidation and reduction reactions. Magnesium appears to be required in order for phosphorescence to occur, although other metals have been shown to be able to perform this task in *in vitro* experiments.

Magnesium binding to ATP

The conversion of sugars to phosphosugars is achieved in the presence of adenosine triphosphate (ATP) and is catalyzed by a kinase enzyme. All kinase enzymes require the presence of divalent metal ions, Mg^{2+} being one of the most common. The magnesium ion complexes with the triphosphate group facilitating cleavage of a phosphorous-oxygen bond.

Manganese in photosynthesis

In photosynthesis electromagnetic radiation, in the form of visible light, is the energy source for a series of redox reactions which can readily be dealt with separately in photosystem I and photosystem II. In photosystem II a strong oxidizing agent is formed and is responsible for the production of molecular oxygen. This oxidizing agent is reduced and recycled for further use by interaction with a manganese complex. During the sequence of events leading to the reduction, manganese displays three of its oxidation states.

Related topics

Phosphoric acid and phosphates (F1)

The early transition metals (G1)

Classification of organic reactions (I2)

Chemical properties

Magnesium is the second member of the Group 2 alkali metals (see Topic A1). All members of this group are reactive and therefore rarely found as the pure metals. When pure magnesium atoms are arranged around each other in a hexagonal pattern, the overall appearance of the metal is silvery. The Group 2 metals are good conductors of heat and electricity. In contrast to the majority of transition metals, magnesium displays only one oxidation state, +2. Magnesium is commonly found as the oxide; this has a melting point of 2800 °C and is an industrially important compound (it is used in the linings of steel furnaces!). Magnesium is also found as a carbonate, a sparingly soluble solid which on heating decomposes to form the oxide with release of carbon dioxide. Magnesium hydroxide, like the oxide, is a very strong base which will neutralize acids and displace ammonia from ammonium salts.

Manganese is a typical transition metal and displays five oxidation states. The highest, +7, is strongly oxidizing, the other states are +6, +4, +3 and +2, the latter being the most stable, and characterized by its pink color in aqueous solution. Pure manganese is a brittle metal with a melting point of 1247 °C. It is of little industrial importance, except in combination with other metals such as iron. Manganese is not readily attacked by air, but will readily combine with oxygen, nitrogen and sulfur on heating, and will liberate hydrogen from warm water. Manganese oxides, particularly of the higher oxidation states, are very powerful oxidizing agents. Although some manganese salts (e.g. the carbonates and phosphates) are water insoluble, the hydroxides of manganese do dissolve in aqueous solution.

Magnesium in photosynthesis

Magnesium appears to perform a variety of roles in biology, not all of them as yet fully understood. However, one of the best characterized systems in which magnesium is involved is chlorophyll (a and b) (see Fig. 1). Chlorophyll is essentially a modified porphyrin ring; one of the carbon linkers between two pyrrole rings is absent making the cavity at the center somewhat smaller than in the model

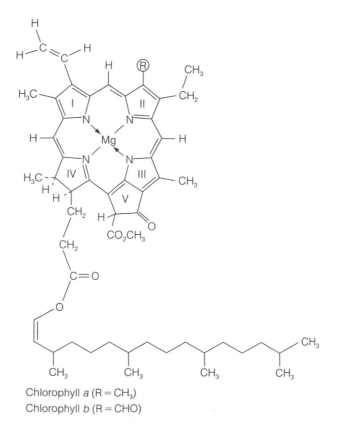

Chlorophyll a (R = CH$_3$)
Chlorophyll b (R = CHO)

Fig. 1.   The structure of chlorophyll a and b.

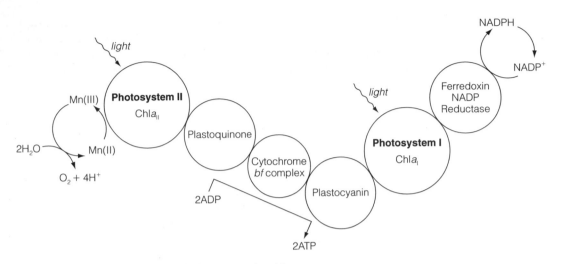

Fig. 2.   Schematic representation of photosystem I and II.

porphyrin (see Topic F1 and *Fig. 1*), but of the correct size to accommodate magnesium ions (amongst other divalent metal ions). The purpose of chlorophyll is to absorb light near the visible region of the electromagnetic spectrum (see Topics Q1 and Q2). In absorbing light chlorophyll then becomes the source of energy for the redox reactions of photosynthesis. There are two general components of photosynthesis; photosystem I and photosystem II (see *Fig. 2*) and chlorophyll is involved in both. When chlorophyll absorbs light this energy can be transmitted into photosystem I to produce a strongly reducing species and a moderately strong oxidizing species. In photosystem II energy transfer leads to an even stronger oxidizing agent (responsible for the generation of molecular oxygen) (*see below*), but weaker reducing agent.

The utility of chlorophyll may be attributed to a number of factors. As the porphyrin system is conjugated, the energy required for electronic transitions is lowered into the visible light region (see Topic Q2). For the system to be conjugated (see Topics L1 and I3) it must be planar and rigid, therefore energy is not wasted in internal motions. In addition chlorophyll is **phosphorescent** (see Topic Q2). This means that an excited state exists for a lifetime of sufficient duration for a chemical reaction to occur, and that energy is maintained in the system during this time. Magnesium, and other divalent metals ions tested in synthetic systems, is vital for phosphorescence. In the absence of the metal only fluorescence occurs and energy is lost almost immediately.

Magnesium binding to ATP

Glycolysis is a metabolic process which is almost universal in biological systems. It involves a sequence of reactions converting glucose into pyruvate with the concomitant production of ATP (see Topics J3 and L2).

The first step in the glycolysis cascade is to phosphorylate glucose, thus trapping it in the cell. Phosphorylation is achieved in the presence of ATP as the phosphate source and a kinase enzyme (hexokinase in the case of this six-carbon sugar). Another requirement is the presence of a divalent metal ion, which is generally magnesium although other divalent metal ions can also perform this role. The magnesium interacts with the oxygens of the triphosphate chain, thereby weakening the phosphorous-oxygen bond and making it more susceptible to cleavage reaction required for phosphoryl transfer (see *Fig. 3*).

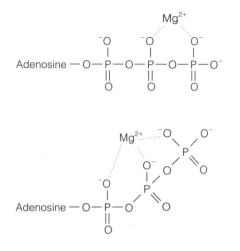

Fig. 3. Proposed structure of complex formed between Mg^{2+} and ATP.

Magnesium has also been implicated in performing a similar role in catalytic systems involving RNA; catalytic RNA systems are referred to as **ribozymes** by analogy to enzymes (see Topics P4 and K1).

Manganese in photosynthesis

Of the two components of photosynthesis, photosystem II (see *Fig. 2*) involves the stronger oxidizing agent. Biologists refer to this oxidizing agent as **P680$^+$**; where P stands for pigment, 680 is the wavelength of light absorbed (this is in the red region of the visible spectrum, other wavelengths are not absorbed so the plant looks 'green'), and the positive sign indicating that this is a cationic species. The 'pigment' is the reactive chlorophyll in a group of chlorophyll molecules described as a photosynthetic unit. The 'nonreactive center' chlorophylls are referred to as **antenna** molecules and simply transfer their energy to a single reactive center chlorophyll. P680$^+$ in conjunction with an enzyme (referred to as Z) extracts electrons from water and leads to the production of oxygen. This water splitting enzyme is a manganese enzyme, containing a cluster of four manganese ions at its catalytic centre (see *Fig. 4*). The sequence of events which lead to the production of oxygen involve manganese changing through a number of oxidation states (see *Fig. 4*). In doing so the manganese system is acting as a **charge accumulator** that enables oxygen to be formed without the generation of hazardous partially reduced intermediates.

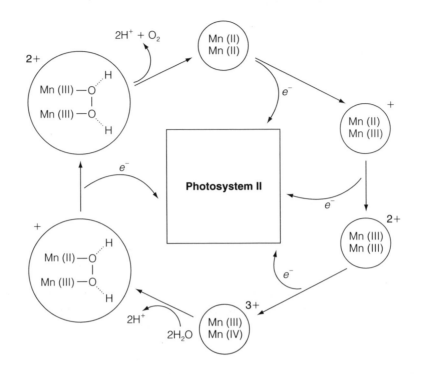

Fig. 4.   Proposed changes in the oxidation state of manganese in photosystem II.

G4 COBALT AND MOLYBDENUM

Key Notes

Chemical properties	Cobalt is a first row transition metal and displays three oxidation states, +2 being the more stable. The main source of cobalt is in ores with arsenic or arsenic and sulfur. Molybdenum is a member of the second series of transition metals and displays four oxidation states. Molybdenum is a trace element but nevertheless is essential for all organisms with the possible exception of green algae.
Cobalt in vitamin B$_{12}$	Vitamin B$_{12}$ is based on a modified porphyrin ring (called a corrin ring) with a cobalt ion at its core. Vitamin B$_{12}$ is the only vitamin known to contain a metal. The complex can undergo a one electron or two electron reduction, the latter leading to a strongly nucleophilic Co(I) species, thought to be key to the methyl transfer reactions of this vitamin.
Molybdenum in xanthine oxidation	Uric acid is the principle end product of purine metabolism across a range of species. The oxidation of xanthine to uric acid is accomplished in the presence of the enzyme xanthine oxidase. The enzyme is a complex ensemble involving two molybdenum atoms, four iron-sulfur clusters, and two flavin adenine dinucleotide (FAD) moieties. Molybdenum utilizes two of its oxidation states in the cascade of reactions involved in the production of uric acid.
Molybdenum in nitrogen fixation	Several bacteria and blue-green algae can fix nitrogen, the enzyme involved in this is called a nitrogenase and varies from species to species. Common to the enzyme systems, which involve two proteins, is the presence of an iron sulfur cluster in the smaller component and an array of two molybdenum atoms, 30 iron atoms and around 30 sulfide ions. Molybdenum has been shown to be vital for activity, replacement of this element by tungsten renders the system inactive.
Related topics	Nitrogen (F2) The early transition metals (G1)

Chemical properties

Cobalt can display three oxidation states, +3, +2, and +1, the latter is extremely unstable, but nevertheless is important in biology. Cobalt is generally found in the form of ores with arsenic or arsenic and sulfur. However, when pure the metal has a bluish-white color and is quite hard, with a melting point of 1490 °C. It is a relatively unreactive element, not readily attacked by air and only oxidized by oxygen at high temperatures. As the +2 oxidation state is the most stable, cobalt (II) compounds are the most common. Cobalt (II) will complex with ammonia and cyanide ions, for example, in solution but on doing so is readily oxidized to Co (III) by these ligands.

Molybdenum is a second row transition metal, in the same group as chromium. It displays four oxidation states, +6, +4, +3, and +2, the first two being very important in biology. Molybdenum is an **essential** trace element; only 1.5 mg of molybdenum is found for every kilogram of the Earth's crust (c.f. silicon, a nonessential element, 282 g per kilogram of earth). One of the isotopes of molybdenum, Mo-98, is one of the most important radioisotopes in diagnostic medicine (see Topic A3).

Cobalt in vitamin B$_{12}$

In 1948 an 'antipernicious anemia factor' was isolated; pernicious anemia is a disease caused by the inability to absorb vitamin B$_{12}$ through the gut wall. This factor is now called **vitamin B$_{12}$** or **cobalamin** (originally the compound was isolated with a cyanide ion associated and was therefore called cyanocobalamin). In 1956 its complex structure was elucidated by Dorothy Hodgkin, using X-ray crystallography. The molecule is built around a corrin ring, that is a porphyrin ring modified by the loss of a pyrrole-bridging carbon (see *Fig. 1*), and cobalt in the middle of the ring in its +3 oxidation state. Cobalt is able to form 6-coordinate complexes and in vitamin B$_{12}$ the fifth and sixth site are occupied by the nitrogen of an imidazole ring and a water molecule. Vitamin B$_{12}$ can undergo one electron or two electron reductions leading to cobalt (II) and cobalt (I), respectively. These reductions may be achieved by nicotinamide adenine dinucleotide (NADH) and flavin adenine dinucleotide (FAD) (see Topic M2);

$$[B_{12}(Co^{III})] \xrightarrow[\text{or FAD}]{\text{NADH (1e)}} [B_{12}(Co^{II})]$$

$$\xrightarrow{\text{(2e)}} [B_{12}(Co^{I})]$$

Vitamin B$_{12}$ in its cobalt (I) form is the substrate for an enzymic reaction which yields a coenzyme, important in the degradation of amino acids (see Topic M1) (see

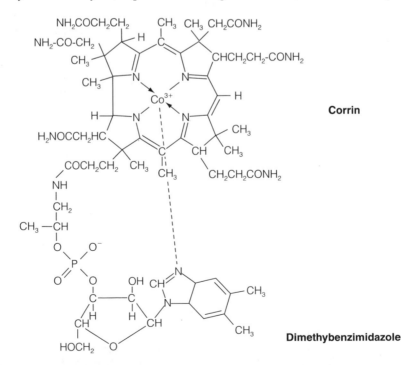

Fig. 1.   The structure of vitamin B$_{12}$.

Fig. 2). Coenzyme vitamin B_{12} is quite unique in biology as it is the only biomolecule involving a carbon-metal bond.

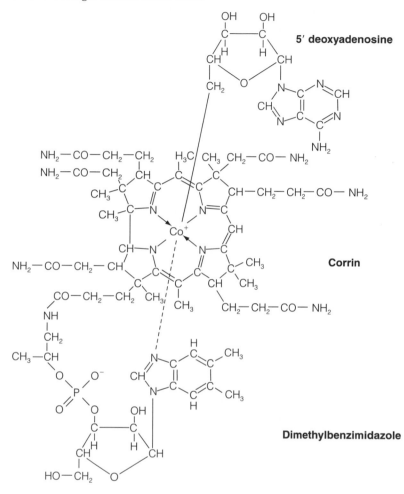

Fig. 2. The structure of coenzyme B_{12}.

Molybdenum in xanthine oxidation

During metabolism the purines, adenine and guanine (see Topics L2 and M2) are broken down to uric acid *via* xanthine. One biosynthetic process involved in the overall metabolic cycle is the oxidation of xanthine. This is achieved with the aid of an enzyme, **xanthine oxidase**. This enzyme is large; its molecular weight is around 300 kDa, and complex, involving two molybdenum atoms, four Fe_2S_2 clusters and two flavin adenine dinucleotide (FAD) molecules. It is believed that one of the intermediates in the oxidation cycle involve molybdenum associated with three sulfur and one oxygen atom;

$$
\begin{array}{ccc}
\text{S—H} & & \text{S} \\
| & -e & \| \\
\text{O}=\text{Mo—S—R} & \underset{+e}{\overset{}{\rightleftharpoons}} & \text{O}=\text{Mo—S—R} \\
| & & | \\
\text{S—R} & & \text{S—R} \\
\text{Reduced} & & \text{Oxidized}
\end{array}
$$

During the course of the oxidation process (see *Fig. 3*), molybdenum converts from the +6 to +4 oxidation state. The overall electron flow may be considered as being from xanthine to molybdenum to Fe_2S_2 to FAD and then to molecular oxygen.

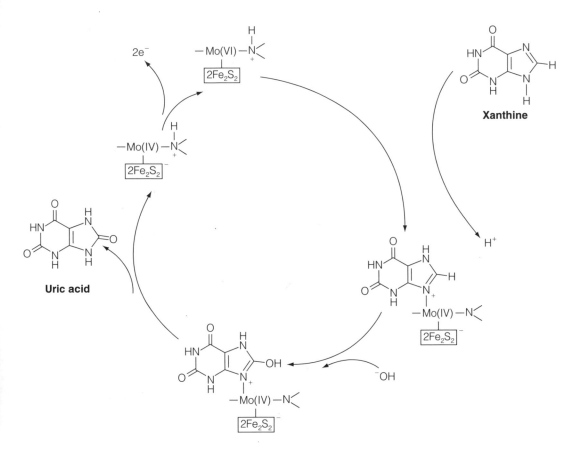

Fig. 3.   Xanthine oxidation.

Molybdenum in nitrogen fixation

Nitrogenase is an enzyme involved in the fixation of nitrogen which occurs in bacteria. This enzyme is comprised of two protein chains. The lower molecular weight protein contains an Fe_4S_4 cluster (see Topic G2). The larger protein, which itself is tetrameric, involves two molybdenum atoms, and large numbers of iron atoms and sulfide ions. Both proteins are required for activity. Although the iron-sulfur clusters are thought to be the redox centers, molybdenum is vitally important (see *Fig. 4*). It has been shown that bacteria grown in the presence of tungsten (VI) oxide rather than molybdenum (VI) oxide, can incorporate tungsten but show no nitrogen fixing activity. It is possible, therefore, that the nitrogen actually coordinates with molybdenum during the fixation process.

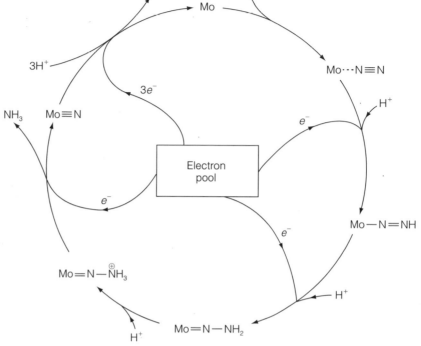

Fig. 4. Proposed catalytic cycle of nitrogen fixation and conversion.

H1 HYDROGEN BONDS

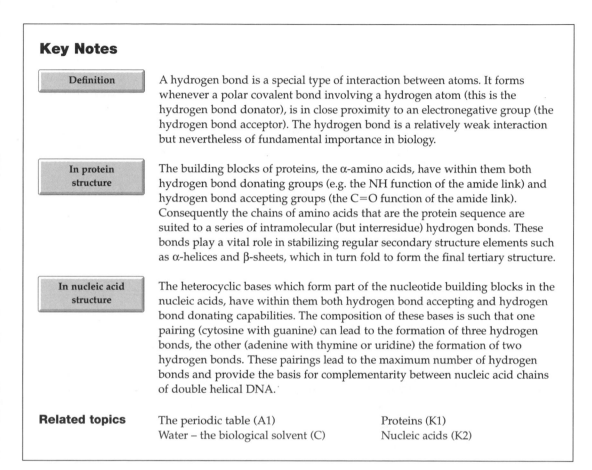

Key Notes

Definition	A hydrogen bond is a special type of interaction between atoms. It forms whenever a polar covalent bond involving a hydrogen atom (this is the hydrogen bond donator), is in close proximity to an electronegative group (the hydrogen bond acceptor). The hydrogen bond is a relatively weak interaction but nevertheless of fundamental importance in biology.
In protein structure	The building blocks of proteins, the α-amino acids, have within them both hydrogen bond donating groups (e.g. the NH function of the amide link) and hydrogen bond accepting groups (the C=O function of the amide link). Consequently the chains of amino acids that are the protein sequence are suited to a series of intramolecular (but interresidue) hydrogen bonds. These bonds play a vital role in stabilizing regular secondary structure elements such as α-helices and β-sheets, which in turn fold to form the final tertiary structure.
In nucleic acid structure	The heterocyclic bases which form part of the nucleotide building blocks in the nucleic acids, have within them both hydrogen bond accepting and hydrogen bond donating capabilities. The composition of these bases is such that one pairing (cytosine with guanine) can lead to the formation of three hydrogen bonds, the other (adenine with thymine or uridine) the formation of two hydrogen bonds. These pairings lead to the maximum number of hydrogen bonds and provide the basis for complementarity between nucleic acid chains of double helical DNA.
Related topics	The periodic table (A1) Proteins (K1) Water – the biological solvent (C) Nucleic acids (K2)

Definition   Atoms in molecules may be held together in a variety of ways. Covalent bonds are a feature of organic compounds (see Topic B2) and ionic interactions are a feature of inorganic compounds (see Topic B2). Covalent bonds are very strong (see *Table 1*) and involve the sharing of valence electrons between atoms of similar electronegativity (see Topic A1). Ionic bonds involve the **electrostatic** interaction between oppositely charged species. In addition, molecules can interact with each other through dipolar attraction. A bond dipole arises when a covalent bond involves atoms of very different electronegativities, such that there is a slight positive charge on one atom of the bond and a slight negative charge on the other atom. Another very important type of interaction is the hydrogen bond. This is a special type of interaction which requires the presence of a hydrogen bond acceptor (an electronegative group, more specifically a group possessing a lone pair of electrons) and a hydrogen bond donor (a hydrogen atom attached to an electronegative atom) (see *Fig. 1*).

Table 1. Energies of interaction between atoms or molecules

Type of interaction	Example	Magnitude (kJ mol^{-1})
Covalent	H-H	200–800
Ionic	Na$^+$ Cl$^-$	40–400
Ion-dipole	Na$^+$ (H$_2$O)$_n$	4–40
Hydrogen bond	H$_2$O $\cdots$ H$_2$O	4–40
Dipole-dipole	SO$_2$ SO$_2$	0.4–4.0

Molecules that have both hydrogen bond accepting and donating capabilities tend to display physical properties characteristic of higher molecular weight systems (see Section C).

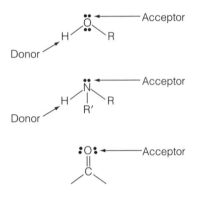

Fig. 1. Hydrogen bond acceptors and donators.

In protein structure

The naturally occurring α-amino acids (see Topic M1) are the basic building blocks of proteins. They combine following a condensation reaction to form amide bonds (see Topic K1);

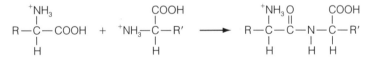

The oxygen atom of the amide carbonyl group is able to act as a hydrogen bond acceptor, whilst the N-H bond of the amide is able to act as a hydrogen bond donator;

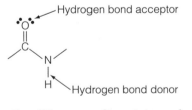

Clearly as these functionalities are adjacent to each other in any particular dipeptide, they are not important in stabilizing the structures of short sequences. However, when longer chains (longer than three residues) are present it is possible that hydrogen bond donation sites and acceptance sites, remote in terms of sequence, can come close in terms of space, with the subsequent formation of hydrogen bonding interactions.

The possibility for hydrogen bonding in peptides and proteins can lead to the formation of regular structural units; referred to as **secondary structure elements**, called **α-helices**, **β-strands**, **β-sheets**, and **β-turns**, amongst others (see *Fig. 2*). These elements are important in providing the scaffolding for the final

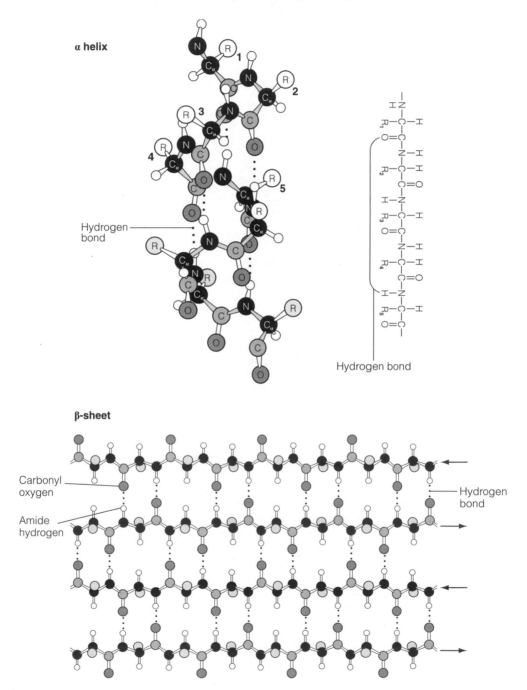

Fig. 2. *Secondary structure elements in proteins stabilized by hydrogen bonding. From: BIOCHEMISTRY by Stryer © 1998, 1995, 1981, 1975 by Lubert Stryer. Used with permission by W.H. Freeman and Company.*

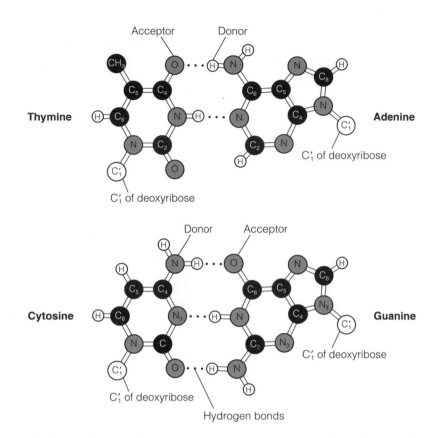

Fig. 3.   Hydrogen bond donator and acceptor sites in natural heterocyclic bases. From:
BIOCHEMISTRY by Stryer © 1998, 1995, 1981, 1975 by Lubert Stryer. Used with
permission by W.H. Freeman and Company.

3-dimensional (or **tertiary**) structure of proteins, which in turn determines the
function of the protein.

**In nucleic acid
structure**

The heterocyclic bases (adenine, guanine, cytosine, thymine and uridine) of the
nucleic acids each have hydrogen bond donating and accepting capabilities (see
Fig. 3). When two oligonucleotide strands (or a folded strand) approach each
other, hydrogen bonding is possible. Indeed hydrogen bonding is key to the
stability of the most common form of DNA, that is the double helix (see *Fig. 4*).

The arrangement of hydrogen bond donor and acceptor sites around the heteroaromatic rings (see Topics L2 and M2), is such that one combination of the four bases leads to a maximum number of hydrogen bonding interactions. This combination requires adenine to interact with thymine (or uridine) *via* two hydrogen bonds and guanine to interact with cytosine *via* three hydrogen bonds. The presence of phosphates and sugars of the nucleic acid means that for these interactions to occur between strands of nucleic acids, the strands must be aligned in an antiparallel fashion (see Topic M2).

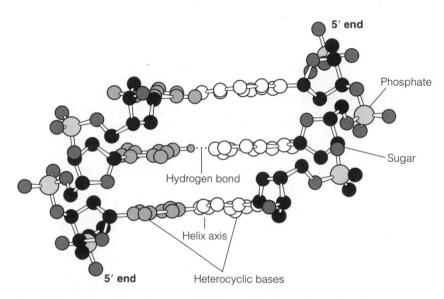

Fig. 4. The double helical structure of DNA is stabilized by hydrogen bonds. From: BIOCHEMISTRY by Stryer © 1998, 1995, 1981, 1975 by Lubert Stryer. Used with permission by W.H. Freeman and Company.

H2 HYDROPHOBIC INTERACTIONS

Key Notes

Definition

The term hydrophobic means water-fearing. Hydrophobic interactions, therefore, are concerned with the establishment of structural units of hydrophobic molecules, which will minimize their contact with water. Consequently hydrophobic interactions may be thought of as passive; hydrophobic molecules are not attracted to each other, they simply repel water.

Importance in membranes

Membranes are made up of lipids and proteins. Lipids are generally linear molecules with a charged, hence hydrophilic (water loving) head group, and a nonpolar, hence hydrophobic tail. Membranes are sheet-like structures comprised of lipid bilayers (lipids meeting tail-to-tail). The driving force for the production of these sheets, which are impermeable to polar molecules (except at gates formed by membrane proteins) is the hydrophobic interactions between arrays of lipids.

Importance in nucleic acids

The most common form of the deoxyribonucleic acids is the double helix. The double helix is formed by two strands of DNA coming together in an antiparallel fashion, enabling the formation of hydrogen bonds between the heterocyclic bases. However, hydrogen bonds cannot fully account for the stability of these molecules. There is an additional stabilizing feature; the hydrophobic interaction between adjacent bases.

Related topics

Hydrogen bonds (H1)   Oligonucleotide synthesis (M2)
Lipids (K3)

Definition

When an oil droplet is suspended in water (see Section C), the hydrogen bonding network between the water molecules is disrupted (see Topics H1 and C) to make a cavity for the **nonpolar**, **hydrophobic** (water fearing) droplet. Water molecules gather around the surface of the droplet, but there is a thermodynamic penalty for doing this. The water molecules **lose entropy** (see Topic O3) hence, this is an unfavorable process. When several droplets are present they tend to cluster together. In forming aggregates some of the water molecules surrounding the surface are released, and there is an overall increase in entropy (for the water) (see *Fig. 1*). A similar process occurs at the atomic level when hydrophobic molecules are involved. In effect the hydrophobic molecules are not interacting as they are not attracting each other, they are being forced together by surrounding water molecules (which have a high affinity for other water molecules).

Hydrophobicity is important in biology as many biomolecules have subunits which possess nonpolar hence, hydrophobic regions. For example, a number of the naturally occurring amino acids (see Topic M1) have hydrocarbon sidechains which are not water soluble. When these are incorporated into a protein the protein must fold such that these residues reside in the interior, away from the surrounding aqueous media.

● = hydrophobic molecule

∧ = water

Large number of water
molecules disrupted

Hydrophobic 'cluster'

Fewer water molecules
disrupted

Fig. 1. Organization of water molecules around one hydrophobic molecule and an aggregate of such molecules.

Importance in membranes

Membranes are highly selective permeable barriers. Cellular membranes regulate the molecular and ionic composition of the cell and also control the flow of information. Membranes are sheet like structures and act as closed boundaries between compartments of different composition. They are comprised of both proteins and lipids. The protein molecules, generally embedded in an array of lipids, act as **gates** and **pumps**, enabling transport across the membrane. The lipid molecules, in lipid bilayers, are the barriers to flow of polar molecules. These bilayers form spontaneously in water due to the hydrophobic nature of the lipid 'tail'. There are a number of different types of lipid (see *Table 1*) but a feature that they have in common is a polar, hence hydrophilic, head group, and a hydrophobic tail (see *Fig. 2*). The most favorable arrangement for lipids in an aqueous

Table 1. Examples of lipids indicating composition of head and tail groups

Lipid	Tail	Head
Phosphoglycerides	Fatty acid chains (long-chain hydrocarbon)	Phosphorylated alcohol
Glycolipid	Fatty acid chain	One or more sugar residues
Cholesterol	Linked carbocyclic systems	'OH' group

(a)

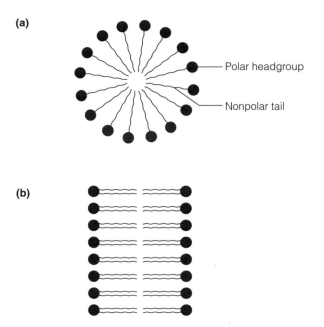

Polar headgroup

Nonpolar tail

(b)

Fig. 2.   Basic lipid structure; micelles and bilayers. (a) Cross-section of a (spherical) micelle. (b) Lipid bilayer showing tail-to-tail alignment.

environment would be with the head groups in contact with water molecules, and the tails away from the water. There are two general ways in which this could be achieved; by the formation of a **micelle** or the formation of a **bilayer** (see *Fig.* 2). Lipid bilayers are of more significance in biology as much larger ordered structures can form than would be possible with micelles. The driving force for bilayer formation is the hydrophobic interactions between the nonpolar tail groups; or rather, the more favorable entropic situation for water around an array of lipids as compared to around individual lipid molecules.

Importance in nucleic acids

In double helical DNA, the two strands of the duplex are held together *via* hydrogen bonding (see Topic H1) between heterocyclic bases on opposite strands. However, hydrogen bonding alone cannot account for the overall stability of the duplex. The presence of some additional stabilizing force could be suggested on the basis of observations made for the bases on their own. When the heterocyclic bases (see Topics L2 and M2), or aromatic molecules (see Topic L1) in general are placed in an aqueous environment, in addition to the formation of hydrogen bonds, resulting in **horizontal stacking** features, the bases are found to **stack vertically** (like coins on top of each other) *(see Fig. 3)*. Several forces seem to be important in this stacking, the lining-up of dipoles, the proximity of pi-electron systems and hydrophobic interactions (i.e. the repulsion of water). Unlike horizontal (hydrogen-bond mediated) stacking, which is purely dependent on nucleic acid composition, vertical base stacking is dependent on nucleic acid composition and sequence.

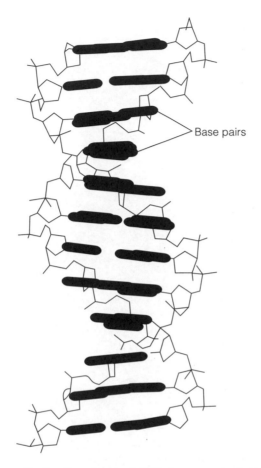

Base pairs

Fig. 3.   Base stacking in DNA (bases are shaded black). Reproduced from Instant Notes in Molecular Biology Second Edition, Turner et al., BIOS Scientific Publishers, 2000.

I1 REACTIVE SPECIES

Key Notes

Nucleophile

The term nucleophile literally means 'nucleus loving', and applies to any species that has an affinity for partially or wholly positively charged centers. Consequently a nucleophile must be electron rich. The location of excess electrons within a molecule is referred to as the nucleophilic center, examples of such are oxygen in water and alcohols, and nitrogen in ammonia and amines.

Electrophile

The term electrophile means 'electron loving', and applies to any species that has an affinity for partially or wholly negatively charged regions of a molecule. Consequently an electrophile is itself electron deficient. An example of an electrophilic center is provided by the carbon atom of carbon oxygen double bonds, or carbon attached to electron withdrawing groups in general.

Free radical

In contrast to nucleophiles and electrophiles, free radicals have no net charge. They are highly reactive, single electron species. Radical reactions in chemical processes generally require the input of large amounts of energy for them to commence. Nevertheless many biochemical processes involve free radicals. For example the reduction of flavin mononucleotide and ribonucleotides in general are radical processes.

Related topics

Classification of organic reactions (I2)   Some metals in biology (G)
Organic compounds by chemical
  class (J)

Nucleophile

When a region of a molecule is **electron rich**, then the molecule is described as being a **nucleophile** (nucleus loving) and the specific region of the molecule is called the **nucleophilic center**. A molecule or region thereof is electron rich if it has a **negative charge**, if it has **lone pairs of electrons** associated with it, or if it involves one or more **pi bonds**. Some examples of nucleophilic species are provided in *Fig. 1*. As nucleophiles, by definition, are electron rich when they react they do so with species that are electron deficient; that is **electrophiles** (see below). In establishing which sites in a molecule are electron rich or electron deficient, considerable progress is made towards predicting the outcome of chemical reactions. When considering the strengths of various nucleophiles, it is informative to take account of the stability of the positively charged species that would arise following nucleophilic attack. Consider the simple example of protonation of hydrogen fluoride (HF), ammonia (NH_3) and water (see *Fig. 2*). As fluorine is the most electronegative element (see Topic A1) possessing three lone pairs of electrons in HF, it would be expected to be the most nucleophilic. In fact in this series it is the least, and this is because fluorine will not tolerate the positive charge that would reside on it. In contrast nitrogen is the least electronegative, but ammonia is the most nucleophilic as it will accept a positive charge, this is

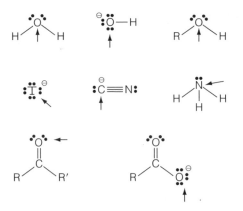

Fig. 1.   Common nucleophilic centers. R, alkyl; R', alkyl or H; ↑, nucleophilic center.

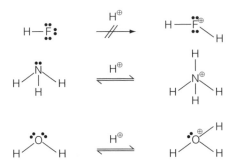

Fig. 2.   Protonation of HF, NH₃ and H₂O. ⇌ , equilibrium.

indicated by the frequency with which ammonium ions appear in chemistry and biology.

Electrophile

The term **electrophile** (electron loving) is applied to atoms or regions of molecules that are **electron deficient**. Electrophiles are, therefore, the exact opposite of nucleophiles, and are the atoms or parts of molecules with which nucleophiles react. Consequently the trend in electrophilic strength is the exact opposite to that of nucleophilic strength. Thus, HF will readily act as an electrophile, taking on negative charge as its hydrogen is abstracted by a base. In contrast NH_2^- which would form if NH_3 was attacked by a nucleophile, is extremely unstable and thus not readily formed. Hence, NH_3 would not be considered electrophilic. In *Fig. 3* some common electrophilic centers are indicated, and in *Fig. 4* the electrophilic and nucleophilic centers for a range of organic compounds (arranged by class) are illustrated.

Free radical

Free radicals are species that possess an **unpaired electron**. In contrast to reactions involving nucleophiles and electrophiles, which are **heterolytic processes** (see Topic B5) (and thus involve the movement of electron pairs), free radicals are involved in **homolytic** processes (see Topic B5), in which there is no net build up or loss of charge.

Free radical processes are highly energetic. A significant amount of energy is required to generate a free radical. Consequently once formed a radical will rapidly

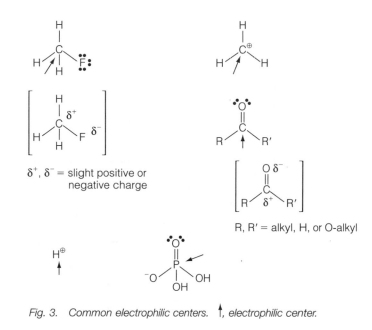

δ^+, δ^- = slight positive or negative charge

R, R' = alkyl, H, or O-alkyl

Fig. 3. Common electrophilic centers. ↑, electrophilic center.

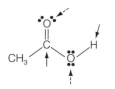

Carboxylic acid

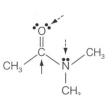

Amide

Anhydride

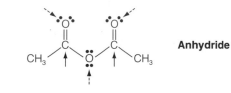

Alkene

Phenyl (aromatic)

Fig. 4. Nucleophilic and electrophilic centers in the various classes of organic compound.
--▶ , nucleophilic center; ─▶ , electrophilic center.

react, often in an apparently nonspecific manner leading to a range of reaction products. To illustrate this reactivity, consider the free radical chlorination of methane which like other hydrocarbons is essentially inert under normal circumstances. The first stage of the reaction requires the input of heat or light to generate chlorine radicals. Once formed these readily react with methane, abstracting a hydrogen radical to form HCl and a methyl radical for further reaction. Several possibilities now arise (see *Fig. 5*) eventually leading to the production of ethane and chloromethane.

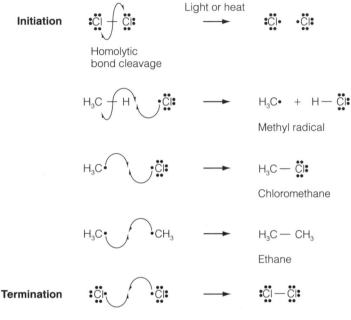

Fig. 5.   *The free radical reaction of chlorine with methane.*

Although free radical processes are extremely energetic, they are crucial in many biological processes (see Topic I3 and Section G). In *Fig. 6* the proposed free radical mechanisms for the reduction of flavin mononucleotide and of ribonucleotides to produce deoxyribonucleotides are shown. The importance of these processes is discussed further in Topic H3.

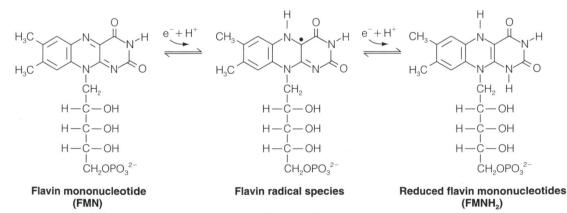

Fig. 6.   *Proposed mechanisms for the reduction of flavin mononucleotide and ribonucleotides in general.*

I2 CLASSIFICATION OF ORGANIC REACTIONS

Key Notes

Addition reactions	Addition reactions, as the name suggests, are reactions which involve the adding of a new chemical group without the loss of an existing group. For an addition reaction to occur the molecule must posses a unit of unsaturation (a double or triple bond).
Substitution reactions	Substitution reactions are those which involve the replacement of part of a molecule (an atom or group of atoms), with an incoming group. In some instances, it is necessary to activate the potential leaving group to make it a better leaving group and thus promote the substitution reaction. Substitution reactions are a feature of carboxylic acid chemistry and aromatic chemistry.
Elimination reactions	Elimination reactions occur with the loss of (generally) a small neutral molecule, from a larger molecule. These reactions are common with alkyl halides and alcohols, leading, in each case, to the formation of alkenes.
Oxidation and reduction processes	In organic systems oxidation and reduction reactions are those which involve the gain or loss of oxygen, respectively. Reduction reactions can also involve the gaining of hydrogen; equivalent to the loss of oxygen. In many instances these processes involve free radicals. Oxidation and reduction reactions are features of the chemistry of almost all organic compounds, and are of great significance in biology.
Related topics	The periodic table (A1) Electron configuration (A2)

Addition reactions

An addition reaction is said to have occurred when a molecule has combined with another, without the loss of any part (atom or groups of atoms) of either molecule. There are two general classes of addition reaction which may be considered, **nucleophilic** and **electrophilic addition** (see Topic I1). For either type of addition reaction to occur one or more **units of unsaturation** must be present in the molecule being added to. Consequently, addition reactions are only possible with, for example, alkenes, alkynes, aldehydes and ketones. Addition reactions are not generally feasible with aromatic compounds as they would result in the loss of aromaticity, a stabilizing force (see Topic L1)

An example of an electrophilic addition reaction is offered by the production of alcohols from alkenes;

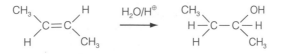

In a similar manner mono- or dihalogenated compounds may be formed;

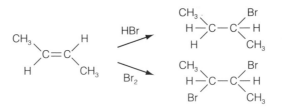

The reactions are termed electrophilic as the carbon-carbon double bond is **electron rich** and therefore attractive to the electrophile (OH_2 or BrH) (see Topic I1). Following the addition of an atom or group to one end of the multiple bond, a positive charge is setup at the other end making this susceptible to nucleophilic attack. Nucleophilic addition is common with carbonyl compounds (see Topic J2) in which the carbonyl carbon is electrophilic and the carbonyl oxygen nucleophilic. Carbon is a better electrophile than oxygen is a nucleophile, and therefore nucleophilic addition is the favored reaction. An example of nucleophilic addition is provided by the production of a cyanohydrin from a ketone and hydrogen cyanide;

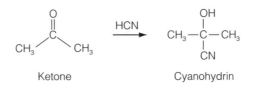

Ketone   Cyanohydrin

Substitution reactions

Substitution reactions are said to have occurred when an atom or group on a molecule is replaced by an atom or group on another. As with addition reactions these may be both nucleophilic and electrophilic; the former being characteristic of alkyl halides, carboxylic acids and derivatives thereof, the latter characteristic of aromatic compounds.

Nucleophilic substitution reactions involving alkyl halides are generally subdivided in terms of the molecularity of the rate determining step for the reaction (see Topic P3). Nucleophilic substitution reactions which involve two species in the rate determining step are referred to as S_N2 reactions; **S** for **substitution**, **N** for **nucleophilic**, and **2** for **bimolecular**. Those involving only one reagent in the rate determining step are referred to as S_N1. An example of each subclass is provided in *Fig. 1*. Whether the pathway followed is S_N1 or S_N2, depends on the leaving halide, but more importantly on the carbon center to which bonds are broken and made. If this carbon can stabilize the buildup of positive charge (leading to a **carbonium ion** or **carbocation**), then the carbon-halogen bond may break before the nucleophile attacks. This would be the **unimolecular** rate determining step. The cation thus formed would have an sp^2 hybridized carbon (see Topic B1), and would therefore have a planar arrangement of substituents about it. This is an important point as if the original alkyl halide had been chiral (see Section E), then a **racemic** mixture (see Section E) would result (see *Fig. 2*). If on the other hand formation of a carbocation was not favorable, then the nucleophile would interact with the carbon before the halogen departed. The rate determining step therefore involves the alkyl halide and the nucleophile and is

(a)

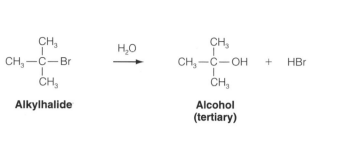

Alkylhalide   Alcohol
                                                               (tertiary)

(b)

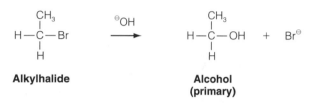

Alkylhalide   Alcohol
                                                               (primary)

Fig. 1.   Examples of (a) S_N1 and (b) S_N2 reactions.

bimolecular. If the alkyl halide was chiral, then the reaction would proceed with **inversion of configuration** (see Section E), as the nucleophile enters from the face opposite the leaving group (see *Fig. 2*).

Aromatic compounds undergo electrophilic substitution reactions; in the case of benzene a catalyst is generally required. These reactions are more favorable than addition reactions as the product retains aromaticity;

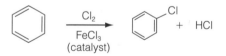

Elimination reactions

Elimination reactions may occur with alkyl halides and with alcohols, resulting in the formation of an alkene and HX (where X is a halogen) or H_2O, respectively. Elimination reactions often compete with nucleophilic substitution reactions, and like those reactions may be subdivided depending on the molecularity of the rate determining step. The label **E2** is given to **bimolecular processes, E1** to **unimolecular** processes.

For an elimination reaction to occur, a good leaving group is required and a strong base. A 'good' leaving group is any group which is stable in its leaving form (e.g. a leaving halide ion is stable due to the electronegativity of the halogens), and hence promotes the forward reaction. If a good leaving group is present, and the carbon to which it is attached can stabilize positive charge then carbonium ion formation will be the first step, and the process will be unimolecular in the rate determining step (E1). Clearly the S_N1 mechanism can compete here. The second step of the reaction involves abstraction of a proton from an adjacent carbon by the base (see *Fig. 3*) to yield the alkene. If, however, carbonium ion formation is not favored, then the first step of the reaction will involve base abstraction of a proton adjacent to the leaving group; the reaction is bimolecular in the rate determining step, E2 (see *Fig. 3*). As many bases can also act as nucleophiles, the S_N2 mechanism can compete here.

S$_N$1

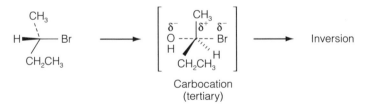

(R)-3-Chloro-3 methyl
hexane

(R)
(retention
of configuration)

(S)
(inversion
of configuration)

Mechanism

Inversion

Carbocation
(tertiary)

Vacant
p-orbital

Retention

S$_N$2

(S)-2-Bromobutane

(R)-2-Butanol

Mechanism

Carbocation
(tertiary)

Inversion

Fig. 2. Mechanism of S$_N$1 and S$_N$2 reactions.

Oxidation and reduction processes

Reduction reactions involve either the loss of oxygen, or the addition of hydrogen and are typical of alkynes, alkenes, aromatics, and carbonyl compounds (see Section I). Oxidation reactions are those which involve the addition of oxygen and are typical of alkenes and aldehydes. These are the definitions routinely used by organic chemists. However, when talking about oxidation and reduction processes in biology, the more general definitions are utilized. That is an atom or compound is **reduced** if it **gains electrons**, and **oxidized** if it **loses electrons**.

(a)

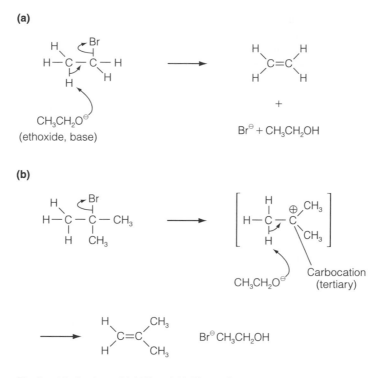

(ethoxide, base)

(b)

Fig. 3.   Mechanism of (a) E1 and (b) E2 reactions.

When, as is usual in biology, oxidation or reduction reactions are reversible then the reactions taken together are referred to as a **redox couple**.

To understand redox reactions in the absence of explicit hydrogen or oxygen molecules, it is first necessary to learn about **oxidation numbers**. The oxidation number of an atom in a molecule, is the charge that the atom would have if all the bonding electrons were owned by the most electronegative atom;

H_2O Oxidation number of oxygen = -2
CH_4 Oxidation number of carbon = -4
NO_2 Oxidation number of nitrogen = $+4$

Clearly, the sum of the oxidation numbers for atoms in a molecule, is equivalent to the overall charge of the molecule. Also, if an oxidation number for an atom becomes more positive during the course of a reaction, an oxidation process is said to have occurred. Conversely, if the oxidation number becomes less positive, a reduction reaction has occurred.

When an oxidation or reduction reaction is reversible, an equilibrium is established with specific rate constants for the forward and backward reactions (see Topics N2 and P1). When writing rate expressions for equilibrium processes, concentrations of reacting species are included. However, redox reactions involve electron transfers and it is difficult to grasp the concept of electron concentration! To solve this dilemma it is useful to talk about **half-reactions**. Every redox reaction has two half reactions; one in which an electron is consumed, the other in which an electron is generated. For a given series of half-reactions, each member would have a distinct potential for 'managing' electrons; this is referred to as the

reduction potential for the reaction in which electrons are consumed. In *Table 1* the **standard reduction potentials** (E^0) for a number of ions and molecules are presented. Standard reduction potentials are the reduction potentials relative to some arbitrary standard, at 1 M concentration, 25°C and 1 atmosphere of pressure for gaseous systems. A **strong reducing** agent has a **negative** reduction potential, a strong oxidizing agent has a positive reduction potential. Knowledge of these reduction potentials enables prediction as to whether a particular combination of reactions will proceed (see *Fig. 4*).

Table 1.   *Some standard reduction potentials (at 25°C)*

Half-reactions		E^0 (volts)
$O_2(g) + 4H^+(aq) + 4e^-$	$\rightleftharpoons 2H_2O$	+1.23
$NO_3^-(aq) + 4H^+(aq) + 3e^-$	$\rightleftharpoons NO(g) + H_2O$	+0.96
$Fe^{3+}(aq) + e^-$	$\rightleftharpoons Fe^{2+}(aq)$	+0.77
$Cu^{2+}(aq) + 2e^-$	$\rightleftharpoons Cu(s)$	+0.34
$2H^+ + 2e^-$	$\rightleftharpoons H_2(g)$	0.00
$Co^{2+}(aq) + 2e^-$	$\rightleftharpoons Co(s)$	−0.28
$2H_2O + 2e^-$	$\rightleftharpoons H_2(g) + 2OH^-(aq)$	−0.83
$Mg^{2+}(aq) + 2e^-$	$\rightleftharpoons Mg(s)$	−2.37
$K^+(aq) + e^-$	$\rightleftharpoons K(s)$	−2.92

Will sodium react with chlorine?

$$Na^+(aq) + e^- \rightleftharpoons Na(s) \qquad E^0 = -2.71V$$

$$Cl_2(g) + 2e^- \rightleftharpoons 2Cl^\ominus(aq) \qquad E^0 = +1.36V$$

As E^0 for chlorine is more positive than for sodium chlorine will be **reduced**, sodium will be **oxidized**

$$Na(s) \longrightarrow Na^+(aq) + e^- \qquad \text{(Oxidation)}$$

$$Cl_2(g) + 2e^- \longrightarrow 2Cl^\ominus(aq) \qquad \text{(Reduction)}$$

(Half-reactions)

Net reaction

$$2Na(s) \longrightarrow 2Na^+(aq) + 2e^-$$

$$Cl_2(g) + 2e^- \longrightarrow 2Cl^-(aq)$$

Sum: $2Na(s) + Cl_2(g) \longrightarrow Na^+(aq) + 2Cl^-(aq)$

Reduction potentials predict this **forward** reaction, but **not** the **reverse** reaction

Fig. 4.   *Example of use of reduction potentials to predict likelihood of reaction.*

A biological example of a series of redox couples is provided by the ubiquinol-cytochrome c system (see *Fig. 5*).

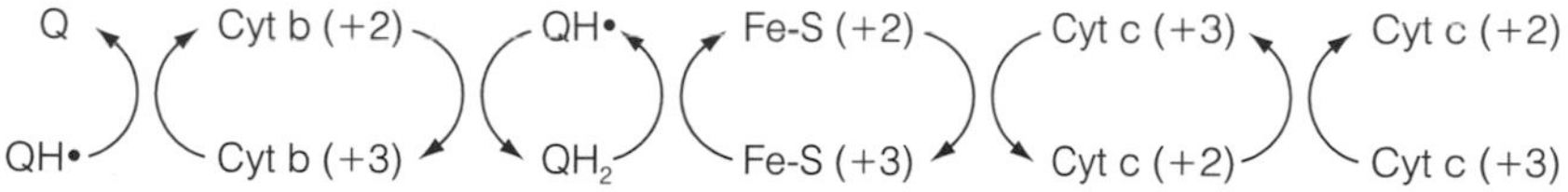

Fig. 5. *Example of a redox couple in biology; the ubiquinol-cyctochrome* c *series. Q, ubiquinone; QH•, semiquinone (radical) intermediate; QH$_2$, ubiquinol; Cyt*b, *cytochrome* b; *Cyt*c, *cytochrome* c; *numbers in parentheses are the oxidation states.*

I3 FACTORS AFFECTING REACTIVITY

Key Notes

Inductive effects	The properties of substituents attached to a system undergoing a chemical reaction (over and above the functional group at which the reaction is taking place), can have a profound effect on the course or feasibility of a reaction. One of the most important substituent properties in this context is that of electron withdrawing power; this is directly related to substituent electronegativity. Electron withdrawing groups are able to stabilize the build-up of negative charge on neighboring atoms. Clearly this is important in, for example, a deprotonation reaction.
Resonance effects	Another electronic property of many substituents is that of resonance. Substituents that have lone pairs of electrons can donate these electrons to form a pi-bond and in doing so stabilize a positive charge. If the substituent is attached to a network of alternate pi-bonds, then resonance effects can be transmitted over all of these bonds *via* delocalization.
Related topics	The periodic table (A1) Aromaticity (L1)

Inductive effects   Several factors affect the course of a reaction. Perhaps the most important of these is the electronegativity of the center which is undergoing change (see Topic H1). However, there are two further factors which contribute to determining the likelihood of a reaction. The first of these is the **inductive effect** of substituents; the second is the ability of substituents to become involved in **resonance**. The importance of substituent inductive effects is readily illustrated by the comparison between the acid-base properties of trifluoroethanol (CF_3CH_2OH) and ethanol. Ethanol has a pKa of 16 (see Topics N2 and N3), and trifluoroethanol a pKa of 12.4. Hence, trifluoroethanol is the more acidic, even though in each case the reaction taking place is the removal of a hydrogen ion (a proton) from an oxygen to leave a negatively charged oxygen (see *Fig. 1*). The reason for the difference lies in the ability of fluorine, the most electronegative element (see Topic A2), to pull electron density towards it, and thereby in effect reduce the electron density of the oxygen and thus stabilize the charged system. If an intermediate or transition state (see Topic E1) of a reaction can be stabilized, it makes the reaction more favorable.

Inductive effects, like that displayed by fluorine (i.e. stabilizing negative charge), are referred to as being negative, those which stabilize a positive charge are referred to as positive inductive effects.

If the hydrogen atom is taken as the reference substituent, with an inductive power of zero, then alkyl groups have a positive inductive effect. It is for this reason that tertiary carbonium ions are more stable than secondary, which in turn are more stable than primary carbonium ions (see Topic I2); an extremely important factor in determining the course of reactions involving carbonium ions.

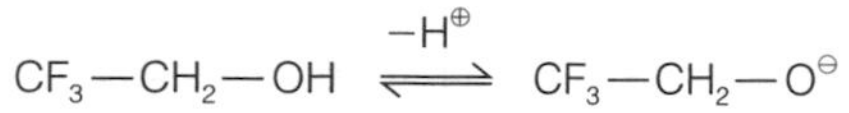

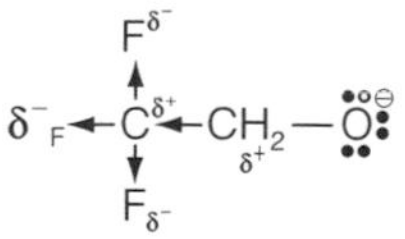

$$CF_3\!-\!CH_2\!-\!OH \;\overset{-H^{\oplus}}{\rightleftharpoons}\; CF_3\!-\!CH_2\!-\!O^{\ominus}$$

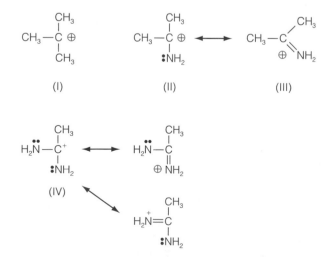

Inductive effect of 'F' stabilizes negative charge,
therefore favors deprotonation (acidity).

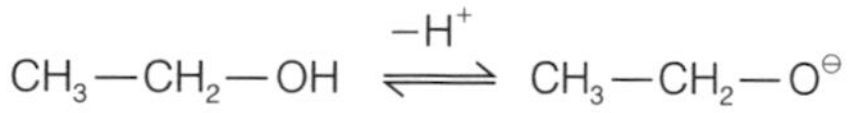

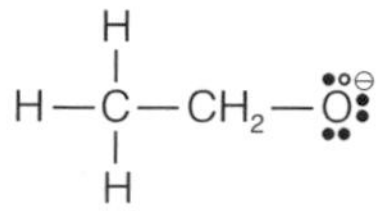

$$CH_3\!-\!CH_2\!-\!OH \;\overset{-H^{+}}{\rightleftharpoons}\; CH_3\!-\!CH_2\!-\!O^{\ominus}$$

No inductive effect to stabilize negative charge.

Fig. 1.   The difference in the pKa of ethanol and trifluoroethanol may be explained by the consideration of inductive effects.

Resonance effects

Inductive effects are of prime importance in reactions in which negative charge builds up. The converse is true of resonance effects. For a substituent to be involved in resonance, which stabilizes a positive charge, it has to be able to give up a lone pair of electrons to enable the formation of a pi-bond. Resonance may be invoked for stabilizing a negative charge if a substituent is able to accept pi electrons to form a lone pair

The significance of resonance may be illustrated by comparing the tertiary carbonium ions (I) and (II) (see *Fig. 2*).

Fig. 2.   Resonance stabilization of a carbonium ion.

As the carbon atom adjacent to the carbonium ion (I) does not possess a lone pair of electrons, resonance is not possible. In contrast, the nitrogen atom attached to the carbonium ion (II) does have a pair of electrons which may be involved in resonance. Indeed the more correct representation for ion (II) is (III); i.e. there are two **canonical forms** (see Topic L1) and the true structure of the reactive species is a **hybrid** of these. The presence of the nitrogen atom leads to the promotion of reactions which involve build-up of positive charge at the attached carbon center.

The more canonical forms that can be drawn for a particular system (charged or otherwise), the more stable the system is (see Topic L1). Hence, carbonium ion (IV) is more stable than ion (III) (see *Fig. 2*).

As inductive effects involve sigma bonding electrons (see Topic B2), these are transmitted over only a few bonds (*Fig. 3*);

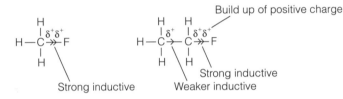

Fig. 3.   Inductive effect reduces over distant sigma bonds.

Resonance effects can, however, be long range. Consider the canonical forms that may be drawn for methoxybenzene (more commonly called anisole) (see *Fig. 4*). One of the oxygen electron lone pairs can become involved with the pi system of the benzene ring, with the result that negative charge builds up on the ortho (2) and para (4) position of the benzene ring. Consequently, these sites on the ring are susceptible to **electrophilic** attack (see Topic I1). The methoxy group (OCH_3) has a positive resonance or **mesomeric** effect and increases the electron density of the ring, making it more reactive. In contrast, an acid function, as in benzoic acid, has a negative mesomeric effect, reducing the electron density of the ring and making it less susceptible to electrophilic attack (see *Fig. 5*).

In *Table 1* the inductive and mesomeric or resonance properties of a range of substituents are listed.

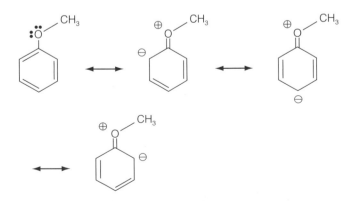

Fig. 4.   The canonical structures of methoxy benzene.

Table 1. The inductive and mesomeric properties of common substituents

Inductive properties	Effect	Group
Strongest electron attracting	−I	—NO$_2$
	−I	—CN
	−I	—F
	−I	—Cl
	−I	—O—C(=O)—CH$_3$
	−I	—C(=O) CH$_3$
	−I	—OCH$_3$
	−I	—N(CH$_3$)$_2$
Electron donating	+I	—alkyl

Mesomeric properties	Effect	Group
Strongest electron donating	+M	—N(CH$_3$)$_2$
	+M	—OH
	+M	—OCH$_3$
	+M	—NH COCH$_3$
	+M	—F
	+M	—Cl
Electron attracting	−M	—COCH$_3$
	−M	—COH
	−M	—CN
	−M	—NO$_2$

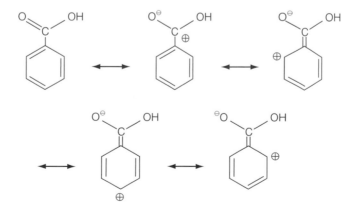

Fig. 5. The canonical structures of benzoic acid.

J1 ALCOHOLS AND RELATED COMPOUNDS

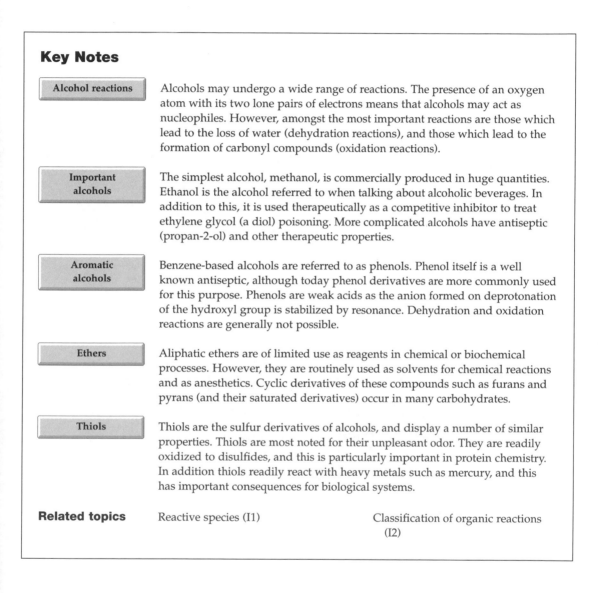

Key Notes

Alcohol reactions

Alcohols may undergo a wide range of reactions. The presence of an oxygen atom with its two lone pairs of electrons means that alcohols may act as nucleophiles. However, amongst the most important reactions are those which lead to the loss of water (dehydration reactions), and those which lead to the formation of carbonyl compounds (oxidation reactions).

Important alcohols

The simplest alcohol, methanol, is commercially produced in huge quantities. Ethanol is the alcohol referred to when talking about alcoholic beverages. In addition to this, it is used therapeutically as a competitive inhibitor to treat ethylene glycol (a diol) poisoning. More complicated alcohols have antiseptic (propan-2-ol) and other therapeutic properties.

Aromatic alcohols

Benzene-based alcohols are referred to as phenols. Phenol itself is a well known antiseptic, although today phenol derivatives are more commonly used for this purpose. Phenols are weak acids as the anion formed on deprotonation of the hydroxyl group is stabilized by resonance. Dehydration and oxidation reactions are generally not possible.

Ethers

Aliphatic ethers are of limited use as reagents in chemical or biochemical processes. However, they are routinely used as solvents for chemical reactions and as anesthetics. Cyclic derivatives of these compounds such as furans and pyrans (and their saturated derivatives) occur in many carbohydrates.

Thiols

Thiols are the sulfur derivatives of alcohols, and display a number of similar properties. Thiols are most noted for their unpleasant odor. They are readily oxidized to disulfides, and this is particularly important in protein chemistry. In addition thiols readily react with heavy metals such as mercury, and this has important consequences for biological systems.

Related topics   Reactive species (I1)   Classification of organic reactions (I2)

Alcohol reactions

Alcohols display a number of properties similar to water (see Section C). They display acid-base behavior and, due to the lone pairs of electrons on the oxygen atom, also act as nucleophiles (see Topics I1 and J3 and Section N). However, some of the most important reactions of alcohols are **dehydration** and **oxidation** processes. The possibility of these reactions taking place is dictated by the exact nature of groups attached to the 'OH' moiety (see *Fig. 1*).

Dehydration reactions are those which result in the loss of water and the formation of an alkene. In the laboratory this may be achieved with ease using an acid such as sulfuric (H$_2$SO$_4$) (see *Fig. 2*). This **elimination reaction** (see Topic I2)

(a)

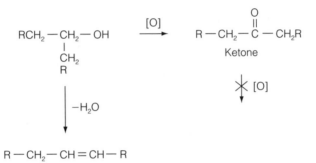

(b)

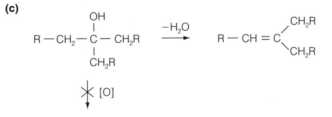

(c)

Fig. 1.   Classification of alcohols and their dehydration and oxidation reactions. (a) Primary alcohol; (b) secondary alcohol; (c) tertiary alcohol.

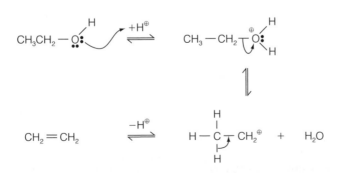

Fig. 2.   The mechanism for the acid catalyzed dehydration of ethanol.

commences with the protonation of the hydroxyl group making it a better leaving group.

Alcohol dehydration is an important reaction in biology. Here the reaction is catalyzed by enzymes (alcohol dehydrogenases) rather than acid. For example citric acid is dehydrated as part of the citric acid cycle;

$$HOOC-\underset{\underset{COOH}{|}}{\overset{\overset{H}{|}}{C}}-\underset{\underset{COOH}{|}}{\overset{\overset{OH}{|}}{C}}-CH_2COOH \longrightarrow \underset{HOOC}{\overset{H}{}}C=C\overset{CH_2COOH}{\underset{COOH}{}} + H_2O$$

The specificity of enzyme catalyzed reactions is such that in general, only one geometric isomer results (see Section E). However, a complication of acid catalyzed dehydration reactions in the laboratory, is the formation of both structural and geometric isomers (see Section E).

Another possible product of an acid catalyzed dehydration reaction is an **ether**. Ethers arise due to the cross reaction between two molecules of alcohol;

$$R-OH + HO-R \xrightarrow[140°C]{H_2SO_4} \underset{Ether}{R-O-R} + H_2O$$

Whether ethers or alkenes are formed may be determined by controlling the temperature within the reaction vessel.

Oxidation reactions are more wide ranging, covering processes which involve either the gaining of oxygen atoms or the loss of hydrogen. Under appropriate conditions alcohols may be oxidized in a controllable manner by agents such as potassium dichromate ($K_2Cr_2O_7$) or potassium permanganate ($KMnO_4$). Oxidizing agents in general may be symbolized by '[O]'.

Primary alcohols are readily oxidized to aldehydes, but further react to form carboxylic acids unless mild oxidizing conditions are utilized (see *Fig. 1*). Secondary alcohols oxidize to form ketones. Tertiary alcohols are generally not susceptible to oxidation processes.

Important alcohols

The simplest alcohol, methanol (CH_3OH) is an extremely important industrial chemical; over 1 billion gallons are produced and used annually. Initially its principal source was from the distillation of wood. Today, however, it is synthetically produced by the reaction of carbon monoxide and hydrogen;

$$CO + 2H_2 \xrightarrow[heat, pressure]{Catalysts} CH_3OH$$

Methanol is a highly toxic alcohol. The consumption of only a few spoonfuls can cause permanent blindness, slightly larger doses can be fatal (see *Table 1*).

In contrast, extending the carbon chain by 1 results in an alcohol which is considerably less toxic than methanol; as the metabolic products of ethanol are less toxic. Ethanol, the alcohol component of all alcoholic beverages, may be produced by fermentation processes, but commercially is bulk produced by the hydration of ethylene;

$$H_2C=CH_2 + H-OH \xrightarrow[300°C]{70\ atm} \underset{H\quad OH}{H_2C-CH_2}$$

Ethanol

Table 1.   Alcohols and their uses

Alcohol	Formula	Uses
Methanol	CH_3OH	Solvent. Precursor of formaldehyde

2-propanol	$CH_3-CH-CH_3$
	  $\quad\quad\quad|$
	  $\quad\quad\quad OH$
		Rubbing alcohol

1,2-propan-diol	$CH_3-CH-CH_2$
	  $\quad\quad\quad|\quad\quad|$
	  $\quad\quad\quad OH\quad OH$
		Moisturizer in lotions and medications

Menthol		In cough drops

Ethanol also has therapeutic uses. Ethylene glycol (CH_2OHCH_2OH), commonly referred to as antifreeze, a diol, is oxidized in the body to form oxalic acid, crystals of which deposit in the kidneys eventually leading to renal failure. However, the enzyme catalyzed conversion of ethylene glycol can be competitively inhibited by the administration of ethanol. The product of ethanol oxidation is ethanoic (acetic) acid, a natural metabolic product (see *Fig. 3*).

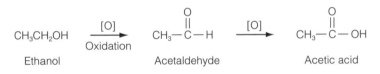

Fig. 3.   Oxidation of ethanol to form acetic acid.

Aromatic alcohols

Aliphatic (saturated) alcohols have the general form ROH, where R is an alkyl chain. Likewise, aromatic alcohols take the general form ArOH, where Ar is a benzene ring. Alcohols of the type ArOH are more commonly referred to as **phenols**, the simplest of which, hydroxybenzene, is almost always called phenol.

Phenols in general exhibit similar properties to aliphatic alcohols; they can act as acids and nucleophiles. However, the proximity of the 'OH' group to the benzene ring precludes oxidation and dehydration processes.

Phenols are widely used as antiseptics; phenol itself was first used as an antiseptic in surgical processes by Joseph Lister in 1800. Substituted phenols are now more routinely used for this purpose;

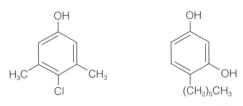

Ethers

Although ethers closely resemble alcohols, the 'H' of the 'OH' group being replaced by 'R', they have little in common with each other. Ethers generally have very low boiling points as they are unable to intermolecularly hydrogen bond (see Topics G1 and C1). For the most part ethers may be thought of as relatively inert, hence their routine use as solvents for chemical reactions. Ethers do, however, have clinical applications as anesthetics;

$$CH_2 = CH - O - CH = CH_2$$
Divinyl ether

$$CH_2F - O - C(F)H - C(F)(Cl)H$$
Enflurane

Of more relevance to biological systems are the cyclic ethers;

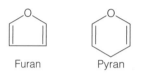

Furan   Pyran

The furan and pyran ring, and hydrogenated analogs thereof, are the basis for many carbohydrates and of course the nucleic acids (see Topic M2).

Thiols

Thiols are noted for their unpleasant odors; these are responsible for the smell of garlic and onions for example. They are the sulfur equivalent of alcohols, and as sulfur and oxygen are in the same group, it would be expected that they displayed similar properties. Thiols can indeed act as nucleophiles and may be oxidized. In contrast to alcohols, however, the principle oxidation product of a thiol is a disulfide;

$$R - SH \quad HS - R \longrightarrow R - S - S - R$$

This reaction is extremely important in protein chemistry (see Topic M1).
The other key reaction of thiols of concern in biology, is that of heavy metal binding;

$$2R - SH + M^{2+} \longrightarrow R - S - M - S - R + 2H^{\oplus}$$

M = heavy metal

Many enzymes contain cysteine with its free thiol group. If such enzymes are exposed to heavy metals then these can tie-up the thiol groups and render the enzyme ineffective, or act as enzyme inhibitors (see Topic P1). A cascade of biological processes could be effected by this.

J2 ALDEHYDES AND KETONES

Key Notes

Physical properties	In general, aldehydes and ketones have lower boiling points than their alcohol counterparts. This is due to there being no possibility for hydrogen bonding between such molecules. However, hydrogen bonds are possible in mixtures of aldehydes or ketones with hydrogen bond donators; such as alcohols and amines. Because of this and the polar nature of the carbonyl bond, small aldehydes and ketones are soluble in water.
Oxidation reactions	Aldehydes can readily be oxidized to carboxylic acids under standard oxidizing conditions. In contrast, ketones are not readily oxidized.
Reduction reactions	Aldehydes and ketones may be reduced by the addition of hydrogen. The products of these reactions are alcohols.
Hemiacetal formations	When an aldehyde reacts with an alcohol, the product of this reaction is referred to as a hemiacetal; this is an α-hydroxy ether. In a similar way, a hemiketal (the product of the ketone and alcohol reaction) may be formed. Both hemiacetals and hemiketals can react with a further mole of alcohol to form acetals or ketals both of which are α-diethers. The cyclic analogs of acetals, ketals, hemiacetals, and hemiketals are of prime importance in carbohydrate chemistry.
Imine formation	Aldehydes and ketones react with amines to form a carbon-nitrogen doubly bonded compound called an imine. This reaction is critical in the production of amino acids.
Keto-enol tautomerism	Carbonyl compounds that have hydrogen atoms on their alpha (carbon) atoms are rapidly interconvertible with their corresponding enols; unsaturated alcohols. This interconversion is referred to as tautomerism. Keto-enol tautomerism is important in determining some of the chemistry of both simple carbonyl compounds and more sophisticated systems, such as the heterocyclic bases of the nucleic acids.
Important aldehydes and ketones	The simplest of aldehydes is formaldehyde, which as an aqueous solution is often used to sterilize surgical instruments. The simplest and most widely used ketone is acetone; its utility arising from its ability to dissolve most organic compounds whilst remaining soluble in water. Aldehyde and ketone functionalities occur in a wide range of biologically important compounds, for example they are present as components of metabolic cycles, sugars, and as hormones.
Related topics	Alcohols and related compounds (J1) Carboxylic acids and esters (J3)

Physical properties

The presence of oxygen in both aldehydes and ketones might lead to the assumption that they would display similar physical properties to alcohols, in particular high boiling points. Alcohols and aldehydes and ketones have oxygen atoms with lone pairs of electrons available for donation in a hydrogen bonding interaction (see Topic H1). However, neither aldehydes nor ketones have groups that may act as hydrogen bond acceptors. If an aldehyde or ketone was to be placed in solution with molecules that can act as hydrogen bond acceptors, then a network of such bonds is established. This ability to hydrogen bond in solution, coupled with the intrinsic polar nature of the carbonyl bond, means that many of the low molecular weight aldehydes and ketones are soluble in water.

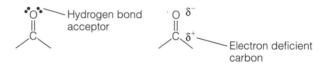

Oxidation reactions

The differing behavior of aldehydes and ketones to oxidizing conditions is the main reason for these compounds being classified separately. Aldehydes are readily oxidized to carboxylic acids;

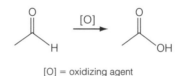

[O] = oxidizing agent

Indeed the reaction is generally so favorable that a sample of aldehyde on standing in air for long periods of time, will generally contain a significant amount of the acid.

A number of chemical tests for the presence of aldehydes have been developed which depend on their ease of oxidation (see *Fig. 1*). Of particular significance to biologists, Benedict's test is often used clinically to establish the presence of glucose in urine samples, a common occurrence with diabetes (see *Fig. 1b*).

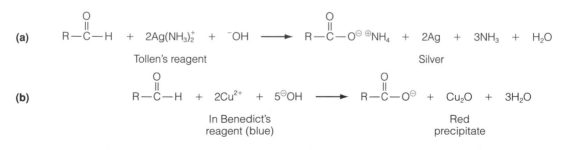

Fig. 1.   Oxidation of aldehydes using (a) Tollen's, and (b) Benedict's reagent.

Under normal conditions ketones are not susceptible to oxidation. Under very forcing conditions in the laboratory it is possible to oxidize ketones (RCOR′) to produce two carboxylic acid molecules (RCOOH and R′COOH).

Reduction reactions

The reduction reactions of aldehydes and ketones are the reverse of the oxidation reactions of alcohols (see Topic J1). These are generally carried out in the laboratory using hydrogen gas and a metal catalyst (see *Fig. 2*).

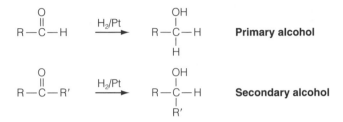

Pt = platinum catalyst

Fig. 2.   Reduction of aldehydes and ketones produces alcohol.

Hemiacetal formation

Aldehydes react with alcohols, generally in the presence of an acid catalyst, to form α-hydroxyethers, commonly referred to as **hemiacetals** (see *Fig. 3*).

Hemiacetals are unstable and cannot usually be isolated. However, if an excess of alcohol is present in the reaction mixture then the hemiacetal reacts further, producing the acetal (α-diethers) which is more stable and isolatable (see *Fig. 3*).

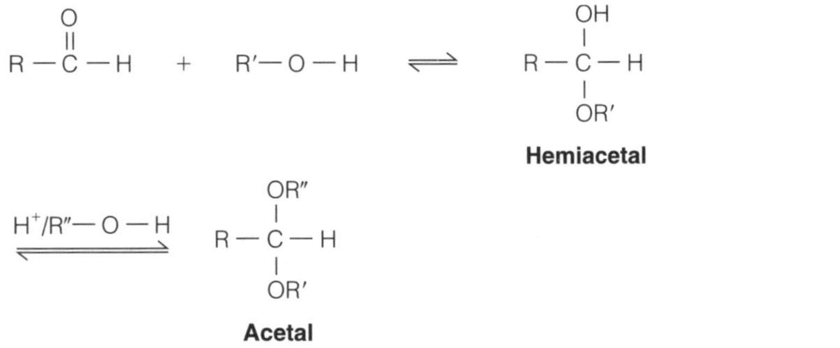

$$R-\overset{\overset{O}{\|}}{C}-H \quad + \quad R'-O-H \quad \rightleftharpoons \quad R-\overset{\overset{OH}{|}}{\underset{\underset{OR'}{|}}{C}}-H$$

Hemiacetal

$$\xrightarrow[\quad]{H^+/R''-O-H} \quad R-\overset{\overset{OR''}{|}}{\underset{\underset{OR'}{|}}{C}}-H$$

Acetal

Fig. 3.   The general reaction of an aldehyde and an alcohol to produce an acetal.

Ketones react in a highly analogous manner to produce hemiketals and ketals. The reactions between ketones or aldehydes and alcohols is of great significance in biology. Both these carbonyl and alcohol functions are present in sugars, leading to the possibility of the formation of cyclic acetals, ketals, etc. For example, glucose, an open chain sugar, spends most of its time in solution as the (intramolecular) hemiacetal, glucopyranose (see *Fig. 4*). These six membered rings are the building blocks for many important disaccharides (see *Fig. 5*) and polysaccharides in general. The five membered ring equivalent derived from ribose, is an important component of nucleic acids.

Imine formation

Aldehydes and ketones, and carbonyl compounds in general, are susceptible to nucleophilic addition reactions (see Section I). Among the most important of these in biology is that in which the nucleophile is ammonia. When ammonia reacts with an aldehyde or ketone, a **carbinolamine** is formed. In the presence of acid a dehydration reaction follows, leading to the production of an **imine**. An identical pathway may be followed with primary amines (NH_2R) also. This reaction is important in biology as it is the key route to introducing amino groups; the reduced imine, into amino acids (see *Fig. 6* and Topic M2).

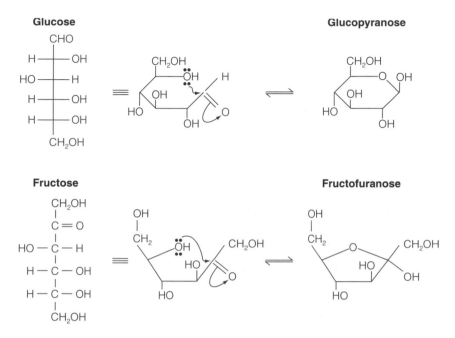

Fig. 4.   *Intramolecular hemiacetal formation: The conversion of glucose to glucopyranose, and fructose to fructofuranose.*

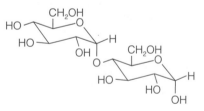

Maltose

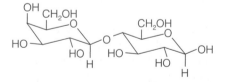

Lactose

Fig. 5.   *Examples of disaccharides, formed by the combination of cyclic hemiactals.*

Keto-enol tautomerism

Aldehydes and, more importantly, ketones that have one or more hydrogen atoms on their alpha carbon, that is the carbon attached to the carbonyl group, are susceptible to a special kind of isomerism (see Topic E1) known as **tautomerism**.

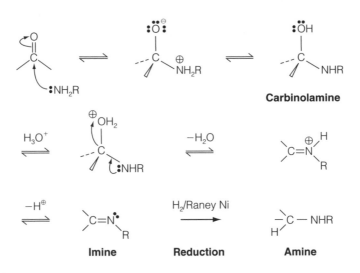

Fig. 6.   The mechanism for imine and subsequent amine production.

This rapid isomerization process is an equilibrium process and involves the formation of an **enol**; that is an unsaturated alcohol. It is therefore referred to as **keto-enol tautomerism**. The position of the equilibrium lies predominantly on the side of the carbonyl compound, enols only being present to a small extent under normal conditions. The position of the equilibrium may however be shifted in the presence of an acid or base catalyst (see *Fig. 7*).

Keto-enol tautomerism is also observed when the atom alpha to the carbonyl moiety is a nitrogen (with an attached hydrogen), as in the heterocyclic bases (see Topic L2). A similar process may also occur with the nitrogen equivalent of a carbonyl, an imine, as has been established for the pyridoxal phosphate catalyzed amino acid transamination process (see Topic L2).

Important aldehydes and ketones

The simplest carbonyl compound is formaldehyde (systematic name methanal), CH_2O. Aqueous solutions of formaldehyde are routinely used to sterilize surgical instruments or to preserve biological specimens. A 37% aqueous solution is referred to as **formalin**.

The simplest ketone is acetone (systematic name propanone). This is extensively used as a solvent due to its ability to dissolve many organic compounds while being soluble in water.

Ketones and aldehydes generally have very pleasant odors and are routinely found to be components of the natural products of fragrant plants (see *Table 1*). They are also present in sugars (see above), and as intermediates in a range of metabolic processes, for example the citric acid cycle (see Topic J1). These carbonyl compounds routinely appear in steroid hormones.

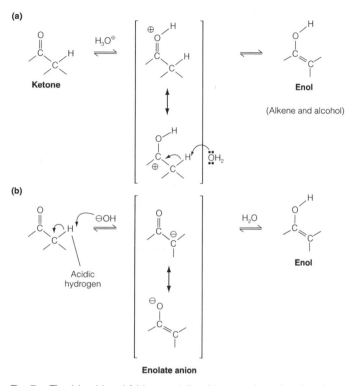

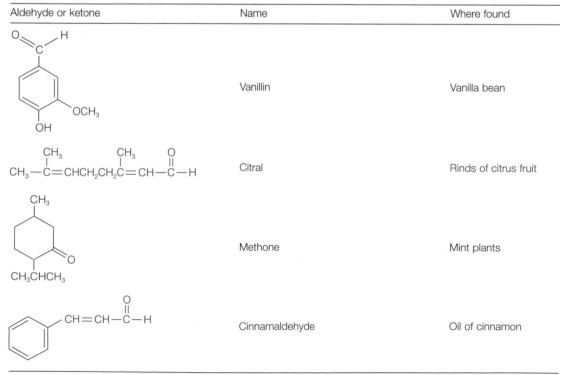

Fig. 7.   The (a) acid and (b) base catalyzed tautomerism of carbonyl compounds.

Table 1.   *Examples of fragrant aldehydes and ketones*

Aldehyde or ketone	Name	Where found
	Vanillin	Vanilla bean
	Citral	Rinds of citrus fruit
	Methone	Mint plants
	Cinnamaldehyde	Oil of cinnamon

J3 CARBOXYLIC ACIDS AND ESTERS

Key Notes

Physical properties of acids	Low molecular weight carboxylic acids are high boiling point liquids at room temperature, generally with unpleasant odors. The reason for the high boiling points is the ability of carboxylic acids to hydrogen bond with like molecules.
Carboxylic acid dissociation	Carboxylic acids are generally quite weak acids, dissociating in water to form the carboxylate ion. In the presence of strong bases such as sodium hydroxide or potassium hydroxide, this dissociation leads to the formation of salts. Salts of carboxylic acids are generally water soluble and are used as preservatives for a range of food products.
Nucleophilic substitution reactions	Like all carbonyl compounds, carboxylic acids are susceptible to nucleophilic attack. One of the most important of such reactions is when the attacking nucleophile is an alcohol. The result of such a reaction is a carboxylic acid ester; usually simply called an ester. When the acid (or derivative thereof) reacts with an amine, an amide is formed. A dehydration reaction between two acids leads to the formation of anhydrides.
Reactions of esters	The most important reaction of esters, both commercially and biologically, is that which involves breaking the ester linkage, leading to either hydrolysis (and acid formation) or saponification (and salt formation).
Esters of inorganic acids	The reaction between acid and alcohol to form an ester is not simply the preserve of organic compounds. Inorganic acids react in a similar manner. The most important inorganic ester in biology is that based on phosphoric acid (H_3PO_4). The phosphoanhydride link is an important component of the nucleic acids. Di- and triester equivalents are found in the key energy sources for biological processes. Their anhydride equivalents, adenosine triphosphate (ATP) and adenosine diphosphate (ADP), are equally important.
Related topics	Phosphoric acid and phosphates (F1) Classification of organic reactions (I2) Reactive species (I1)

Physical properties of acids

The simplest carboxylic acids are generally liquids at room temperature with characteristic, and generally unpleasant, odors (see *Table 1*). However, they are high boiling point liquids. As with alcohols (see Topics I1 and J1) carboxylic acids are able to intermolecularly hydrogen bond. Carboxylic acids generally have higher boiling points than alcohols of the same molecular weight, as hydrogen bonded acids are held together by two bonds;

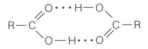

This ability to hydrogen bond makes the smaller carboxylic acids water soluble.

Table 1. Examples of naturally occuring carboxylic acids

Carboxylic acid	Name	Where found
$CH_3(CH_2)_{12}CO_2H$	Myristic acid	Nutmeg
$C_6H_5CH = CH\ CO_2H$	Cinnamic acid	Cinnamon

CH_3CHCO_2H
   |
   OH
	Lactic acid	Sour milk

Carboxylic acid dissociation

Perhaps the most important property of carboxylic acids is their acidic behavior (see Section N). In water, a proton leaves the acid, converting it to a carboxylate anion (*Fig. 1a*). Carboxylic acids are relatively weak acids and therefore are only fractionally dissociated under normal aqueous conditions. However, when strong bases, such as sodium hydroxide or potassium hydroxide are present, carboxylic acids readily dissociate to form salts (*Fig. 1b*).

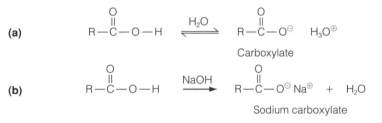

Fig. 1. *Carboxylic acids dissociate to small extent in (a) water but to greater extent in (b) the presence of a strong base.*

These salts are generally solid at room temperature and water soluble. A number of carboxylic acid salts are used commercially as food preservatives. For example, calcium and sodium proprionate ($CH_3CH_2COO^-\ Na^+$) are used in the preservation of breads and cakes. Sodium benzoate ($C_6H_5COO^-\ Na^+$) occurs naturally in many foods but is also added to a range of food stuffs.

Nucleophilic substitution reactions

Carboxylic acids, like all compounds containing carbonyl groups, are susceptible to nucleophilic attack (see Section I) due to the slight positive charge on the carbon of this functionality. In terms of biological processes, amongst the most important of such reactions are those which lead to the formation of carboxylic acid esters, amides (see Topic M) and anhydrides.

As long ago as 1895, Fischer and Speier discovered that esters result from simply heating a carboxylic acid in an alcohol solution containing a small amount of mineral acid. This reaction is now commonly referred to as the **Fischer esterification**, and its mechanism is well understood (see *Fig. 2*).

The conversion of carboxylic acids to amides is not straight forward. The simple addition of an amine tends to lead to salt formation;

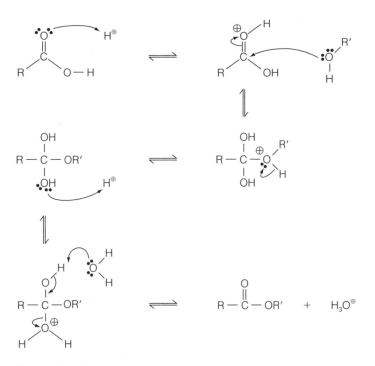

Fig. 2.   The Fischer ester synthesis.

The reaction needs to proceed by some intermediate, usually an acid chloride, then the reaction is simply nucleophilic displacement of the chlorine by the amine (see Topic M1). Carboxylic acid anhydrides also feature in biochemistry and these are formed by the combination of two molecules of acid and the loss of a water molecule (see *Fig. 3*).

Reactions of esters

Both commercially and in biology the most important reaction of esters is that which involves cleavage of the ester linkage, that is the C-O single bond. Such a process occurs with hydrolysis and with **saponification**. Ester hydrolysis can readily take place in the presence of an acid catalyst (see *Fig. 4*). The mechanism for this reaction has been established by the use of ^{18}O isotopically enriched water (see Topic A3); the label is found to reside on the ester at the end of the reaction. Ester formation and hydrolysis are fundamental reactions in biology, where the catalyst is of course enzymic. Animal fats and vegetable oils are esters and hydrolysis is an important process by which they are digested.

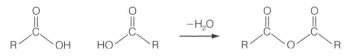

Fig. 3.   Carboxylic acids undergo a dehydration reaction to form acid anhydrides.

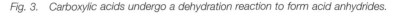

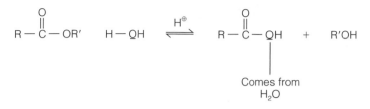

Fig. 4.   Ester hydrolysis.

Saponification is another ester-linkage breaking process, but takes place in the presence of strong bases (cf. carboxylic acid salt formation) (*Fig. 5*);

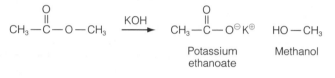

Fig. 5.   Ester saponification.

The saponification of fats and oils is important in the production of soaps.

Esters of inorganic acids

Carboxylic acid based esters are extremely important in biology, however other types of esters are equally important. Just as carboxylic acids react with alcohols to form the corresponding esters, inorganic acids can react to form inorganic esters, the most important of these in biology are those from phosphoric acid (see Topic F1);

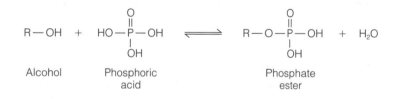

As phosphoric acid has three 'OH' groups it is able to form di- and triesters (the linkages also referred to as anhydrides);

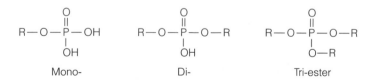

Mono- and diesters are essential to life and represent some of the most important biological molecules. For example, the monophosphate of glucose is the first intermediate in the oxidation of glucose, a key energy source for the body.

Like carboxylic acids, phosphoric acid can form dimers and trimers by a dehydration process. These anhydrides are referred to as **diphosphate** and **triphosphate esters** (or anhydrides), the phospho components of the key energy sources adenosine triphosphate and adenosine diphosphate (see Topics F1 and G3 and M2).

J4 AMINES AND AMIDES

Key Notes

Physical properties of amines and amides	The simplest amines are gases at room temperature; as molecular weight increases liquids and solids are also found. The lone pair of electrons of the nitrogen atom enable amines to act as hydrogen bond acceptors; primary and secondary amines can also act as hydrogen bond donors.
Reactions of amines	Amines act as weak bases (and nucleophiles) and readily form salts when in solution with acids. Amine salts are soluble in water, hence amine containing drugs are often administered as the salt. One of the most important reactions of amines is that with carboxylic acid derivatives to form amides.
Reactions of amides	One of the most important reactions of amides is that of hydrolysis to form a carboxylic acid and to generate an amine salt, or carboxylic acid salt and amine, depending on reaction conditions.
Biologically important amines and amides	A number of naturally occuring compounds have the amine group at their core. Many neurotransmitters, amphetamines and alkaloids are amine based. Similarly a number of pharmaceutical products are amide based; for example diazepam and ampicillin.
Related topics	Classification of organic reactions (J2) Peptide synthesis (M1)

Physical properties of amines and amides

The simplest amines are gases at room temperature. As the amine becomes more substituted the boiling point increases and liquids and solids are possible. In contrast, the simplest amide, formamide ($HC(O)NH_2$), is a liquid at room temperature. As substitution increases solids are more common.

Both amines and amides are capable of hydrogen bonding, and this has a predictable effect on boiling points (see *Fig. 1*).

Amines are characterized by their unpleasant odors; low molecular weight amines smell like ammonia.

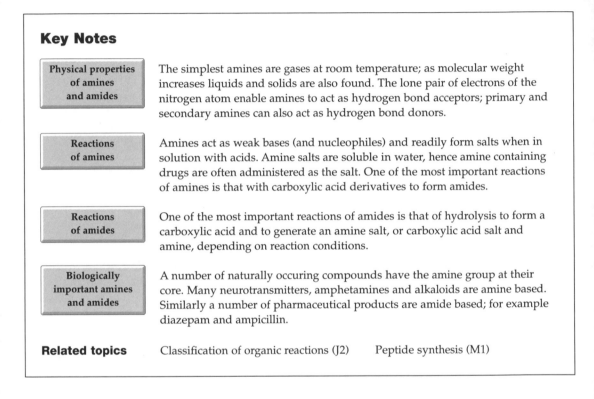

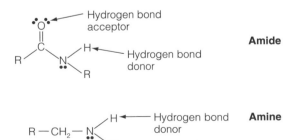

Fig. 1.   Hydrogen bonding capabilities of amines and amides.

Reactions of amines

Amines, like alcohols and water, display basic properties (see Section N); indeed amines are the most common organic bases. Amines readily accept a proton from water to form ammonium ions. If, however, an acid is present then an amine salt is produced.

Amine salts, like carboxylic acid salts (see Topic J3), are usually water soluble. As the amine group is a common feature of drug molecules, drugs are often administered as the amine salt to ensure that the drug is absorbed;

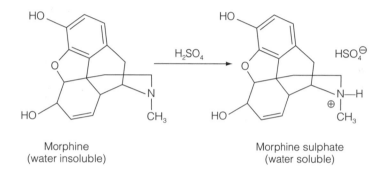

Morphine   Morphine sulphate
(water insoluble)   (water soluble)

Perhaps the most important reaction involving amines is that which results in amide formation (see Topic J3) (*Fig. 2*)

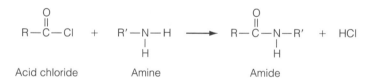

Acid chloride   Amine   Amide

Fig. 2. Amide formation.

Reactions of amides

One of the most important things to note about amides in comparison to amines is their neutrality. The lone pair of electrons on the nitrogen atom of the amide is not available for protonation to take place. This lone pair is actually delocalized with the pi electrons of the C=O double bond (see Topic I3). This feature has been established by the determination of the C-N bond length, which is shorter than a single bond, and helps explain why free rotation about the C-N bond is not possible. The most important reaction of amides is that of hydrolysis; this is the reverse reaction of amide formation (see Topic J3). The product of hydrolysis depends on the reaction conditions utilized. The acid catalyzed reaction results in the formation of an amine salt. The base catalyzed reaction results in the formation of an amine (see *Fig. 3*).

Amide hydrolysis is very important in biology. It is a central reaction in the digestion of proteins.

(a)

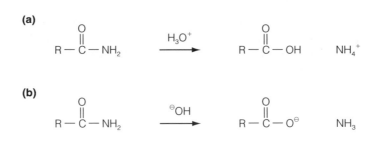

(b)

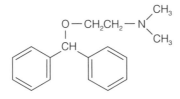

Fig. 3.　The dependence of reaction conditions on the products of amide hydrolysis. (a) Acid catalyzed; (b) base catalyzed.

Biologically important amines and amides

Many naturally occurring compounds and pharmaceutical agents contain amine or amide functions. Among the most important amine compounds are the neurotransmitters, such as dopamine and serotonin (see *Fig. 4*). These play an essential role in the transmission of nerve impulses throughout the body. Amphetamines, such as benzedrine, are generally powerful stimulators of the

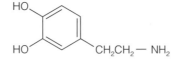

Diphenylhydramine
(Benadryl, an antihistamine)

Dopamine
(a neurotransmitter)

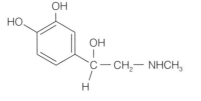

Adrenaline
(stimulant of central
nervous system)

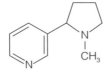

Nicotine
(stimulant)

Fig. 4.　Examples of biologically important amines.

nervous system. Although many of these are used legally as drugs, they are also used illegally to heighten or stimulate emotions (see *Fig. 4*).

Alkaloids are amine based compounds generally isolated from plants, and amongst the most powerful psychotropic drugs. Some of these drugs are used to cure diseases others, such as nicotine, cause them and are addictive (see *Fig. 4*).

In addition to amides being the key structural and functional unit of peptides (see Topic M1), they are the core of numerous medicinal products (see *Fig. 5*). For example, acetaminophen, marketed under a number of names including Panadol, is an effective alternative to aspirin, without the side effects of gastrointestinal bleeding! However, acetaminophen has no anti-inflammatory activity, therefore unlike aspirin cannot be used for the treatment of rheumatoid arthritis.

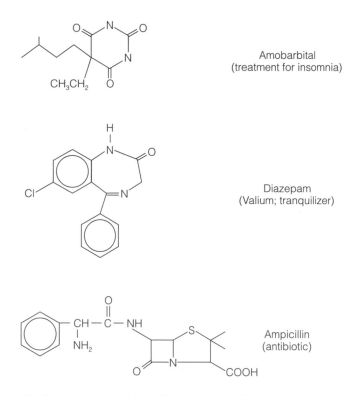

Amobarbital
(treatment for insomnia)

Diazepam
(Valium; tranquilizer)

Ampicillin
(antibiotic)

Fig. 5.   Examples of biologically important amides.

K1 PROTEINS

Key Notes

Role of proteins

Proteins perform a variety of functions. Proteins can catalyse chemical reactions, facilitate the transport and storage of small molecules, mediate an immune response and control cell growth, amongst other roles. Proteins are involved in almost every biological process.

Composition of proteins

Proteins are polymeric molecules. The monomer precursors of these large molecules are the L-α-amino acids. The polymer is formed by a condensation reaction between two amino acid residues with the direction of synthesis being controlled.

Protein structure

Protein molecules can form very complicated structures and these structures determine their function. The primary structure of any protein is the sequence of amino acids together with information regarding any disulphide bonds that may be present. The secondary structure of a protein refers to the folding of the polymer chain to form regular or ordered structural motifs such as α-helices or β-strands. Tertiary structure refers to the global fold of the molecule and the organization of secondary structure units. A number of important proteins are comprised of more than one polypeptide chain, each chain called a subunit. The way in which different subunits interact is referred to as the quaternary structure of the molecule.

Related topics
Peptide synthesis (M1)   Enzyme kinetics (P4)
Hydrogen bonds (H1)

Role of proteins

Proteins are very versatile macromolecules involved in almost all biological processes. Proteins can provide the impetus (or act as **catalysts**) for chemical reactions to occur under physiological conditions. In many instances a protein can increase the rate of a reaction a million-fold. In this role proteins are referred to as **enzymes** (but not all biological catalysts are proteins). Protein molecules can act as the carriers of important small molecules or ions. For example, the protein hemoglobin transports a molecule of oxygen (see Topic F3 and G3) in erythrocytes and iron is transported by proteins called transferrins in blood plasma (see Topic G2). **Antibodies** are proteins which recognize foreign agents such as bacteria, and group together to remove them. The mechanical properties of proteins in, for example, our skin, is due to the mechanical properties of a fibrous protein called **collagen**. Proteins are involved in the generation and transmission of impulses to nerve cells and they are involved in cell growth and differentiation. In the latter role the proteins are highly specialized molecules called **hormones** (see Topic K3). An example of a hormone that is also a protein is **insulin**.

Composition of proteins

Proteins are polymers comprised of amino acid monomers (see Topics M1 and H1). There are some 20 amino acids. These are all α-amino acids and all but one of them (glycine) is chiral (see Topics E2 and E3). The configuration of the naturally occurring α-amino acids is L (see Topic E2). The amino acids differ in the nature of their sidechain 'R':

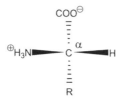

The sidechain may be polar, non-polar, positively charged or negatively charged (see Topic M1). The properties of 'R' dictate the structure of the protein and hence its function (see below). The polymerization process involves the 'condensation' (loss of water) of a free amino group of one amino acid and a free carboxyl group of another to form a peptide bond (see Topic M1) which generally has a *trans*, planar geometry (see Topic E2):

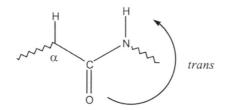

The polymerization process occurs such that the first amino acid in the final protein will have a free amino group (or N-terminal) and the last, a free carboxyl (or C-terminal) group.

Protein structure

Proteins are long chain molecules that in some instances contain several hundred amino acids, it is not surprising therefore that they do not generally exist in an extended chain form. Rather these polymers fold. Polypeptide chains fold in a variety of ways and in doing so enable favorable interactions between backbone (amide) or sidechain atoms and remove unfavorable interactions, such as those between like charges or hydrophobic sidechains (see Topics H1 and H2) and water, the solvent. Thus, in describing a protein it is not sufficient to simply know the amino acid composition; the sequence, which dictates the folding, must also be known. The sequence of the amino acids, with residue '1' the N-terminal residue (see above), is referred to as the **primary structure**. From this, together with knowledge of the 'state' of cysteine residues (see Topic M1) it is possible to predict, with varying degrees of success, where folds in the molecule may occur i.e. the location of **secondary structure** elements (see Topic H1). Some amino acids sequences are prone to fold to form α-helices (for example, those containing methionine or alanine, see Topic H1) others to form β-strands, β-sheets (those containing valine or isoleucine) or β-turns (those containing proline). In between these regular structural motifs there are often regions with more flexibility enabling secondary structure motifs to come into contact with each other to produce the **tertiary structure** of the polymer. By definition therefore the tertiary structure refers to the spatial arrangement of amino acid residues which are well separated in terms of sequence position. In forming this 'final' structure a variety of interactions are enabled. These include hydrogen bonding, hydrophobic, and electrostatic (and covalent in the case of a disulphide bridge) interactions. It is not until the protein is fully folded in this manner that the molecule is in its active form. If a protein involves more than one polypeptide chain (each of which folds to a tertiary structure) then each chain is referred to as a subunit. Hemoglobin has four subunits, the description of how these four subunits interact is referred to as the **quaternary structure**.

The forces holding the folded protein together can easily be disrupted, for example by heat, or changing the solvent. If a protein sample is dissolved in chloroform rather than water, the hydrophobic interactions which define a tertiary structure are 'switched off' and the protein may unfold. Unfolded proteins with their extended backbones and exposed sidechains often clump together in aggregates which may come out of solution. This is what happens to egg albumin when exposed to heat (the hard egg white is formed).

K2 NUCLEIC ACIDS

Key Notes

Role of nucleic acids	There are two types of nucleic acids; the deoxyribose (DNA) and the ribose (RNA) forms. DNA carries the genetic code for life in cells. DNA acts as its own template for the production of further DNA molecules (replication). RNA is pivotal in the interpretation of the genetic code carried by the DNA and in the control of protein synthesis (transcription and translation). RNA is also the genetic material for a large number of viruses.
Composition of nucleic acids	Nucleic acid molecules are polymers. The monomer precursors of these high molecular weight systems are nucleotide triphosphates (NTP). NTPs differ in the nature of the sugar ring (deoxyribose in DNA ribose in RNA) and the heterocyclic bases (adenine (A), guanine (G), thymine (T) and cytosine (C) in DNA, A, G, C, and uracil (U) in RNA).
Nucleic acid structure	DNA is most commonly found in a bi-molecular complex with two polymer strands interacting in an anti-parallel orientation. The strands are held together by hydrogen bonds and the bi-molecular system, or duplex, is further stabilized by pi-stacking of the heterocyclic bases. The most common shape of the duplex is a right-handed helix as first reported by Watson and Crick. More complicated folding patterns for DNA are emerging. RNA molecules can fold to form duplexes but are also found in single stranded form.
Related topics	Oligonucleotide synthesis (M2), Hydrophobic interactions (H2) Hydrogen bonds (H1),

Role of nucleic acids

Until 1944 it had been assumed that our genetic make-up was dictated by proteins (chromosomal proteins) (see Topic K1) and that nucleic acids (more specifically deoxynucleic acids, DNA) played a secondary and thus less important role. This view was dispelled by Oswald Avery and co-workers in their work on *Pneumococcus*, which showed that DNA in the absence of proteins had genetic specificity. It is now widely accepted that DNA carries the genetic code for all cells. Subsequently the structure of DNA molecules was determined (see below) and the mechanisms of its function elucidated. The structure suggested a simple process for directing the synthesis of new DNA molecules; the original, or **parent** DNA acting as a template for **daughter** molecules (see *Fig. 1*). This is **replication**. The specificity of the interactions between heterocyclic bases drives the synthesis of the correct sequence; the process is catalyzed by DNA polymerases (see Topic K1).

The sequence of the bases in DNA (our **genes**) determines the proteins which are produced by various cells (see Topic K1), however DNA is not directly involved in protein synthesis. The instructions for protein synthesis are carried out by RNA molecules which are in three basic categories; **messenger (m)**,

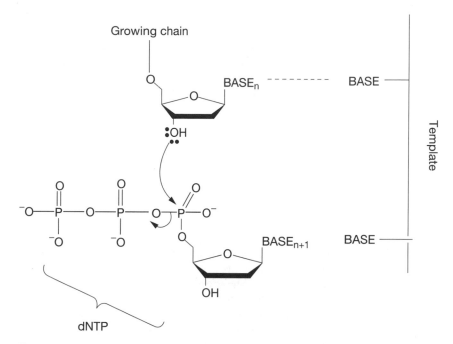

Fig. 1. *Template directed synthesis of DNA.*

transfer (t), and ribosomal (r), giving rise to mRNA, tRNA and rRNA. Under the control of RNA polymerases an RNA complement of the DNA template is made. This process is called **transcription** and results in the formation of mRNA. At a part of the cell called the **ribosome,** tRNA (carrying an appropriate amino acid) and rRNA follow the instructions of the mRNA to **translate** the nucleic acid sequence into an amino acid sequence.

Although we have stated above that DNA contains the genetic information in cells, a wide range of viruses use RNA as their genetic material. These viruses include HIV, measles, influenza, foot and mouth, ebola and polio. Their life cycles can be complicated and may not necessarily involve DNA.

Composition of nucleic acids

Nucleic acids are high molecular weight molecules composed of, in some instances, many thousands of monomer units. The monomer units are called **nucleotides** (see Topic M2) which occur naturally as triphosphate analogs (see *Fig. 1*). Nucleotide triphosphates differ in the sugar ring; ribo-NTPs (rNTP) have a ribose sugar, deoxyribo-NTPs (dNTP) have a 2′-deoxyribose sugar (see Topic M2), and in the base attached to this ring; adenine (A), guanine (G), thymine (T), and cytosine (C) in dNTPs, with thymine being replaced by uracil (U) in rNTPs. The NTPs are attached to each other in a specific direction such that the last nucleotide in any given sequence will have a free 3′-OH and a 5′-phosphate group. In tRNAs and rRNA, modified bases (called 'minor bases') are sometimes encountered.

Nucleic acid structure

Nucleic acids are long chain molecules which are often found associated with a **complementary** long chain molecule; a bi-molecular complex. The sequence of a nucleic acid molecule has directionality (cf. N-terminus and C-terminus of proteins (Topic K1)); that is a molecule of sequence d(ACGT) will have a free 3′-OH on the thymine residue and a phosphate (or triphosphate) on the 5′-position of the adenine residue. Thus, the sequence could be written 5′-d(ACGT)3′; the complement to this strand being 3′-d(TGCA) 5′. These two strands may interact by the formation of hydrogen bonds (see Topic H1); A with T (or U in the RNA system) and G with C, in an anti-parallel orientation;

```
      ────────────────────►
5′   A     C     G     T    3′
3′   T     G     C     A    5′
      ◄────────────────────
```

When two strands interact a double helix is formed, this is often called the **Watson-Crick** double helix after the two scientists who in 1953 famously first determined the structure of DNA (see Topic H2). The double helix, or duplex, is further stabilized by favorable pi-electron interactions between the heterocyclic bases along the vertical axis (see Topic H2). More recently DNA has also been found to adopt more complex structures such as triplexes, in which a third DNA strand wraps around the major groove of the double helix (see *Fig. 2*), and tetraplexes. The physiological relevance of these structures is still under investigation.

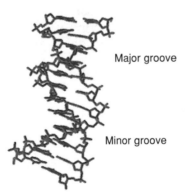

Major groove

Minor groove

Fig. 2.   Double helix commonly found for DNA molecules.

RNA molecules are able to interact with other RNA molecules to form duplexes similar to those observed for DNA. A more common feature of RNAs is the folding of single strands to form loops and bulges, interspersed by stretches of intramolecular double-helix (see *Fig. 3*).

Such loops are a feature of tRNAs in which a single RNA chain folds to form a structure which, when drawn to illustrate the pattern of the base-pairing, resembles a clover-leaf (see *Fig. 4*).

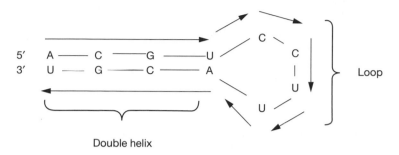

Fig. 3. An RNA sequence folded to form a 'loop' and a double helix.

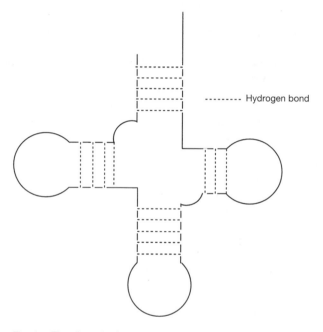

Fig. 4. The clover-leaf structure of t-RNA.

This is a representation of the nucleic acid **secondary structure**. The three-dimensional structure was found to be like the letter 'L'.

K3 LIPIDS

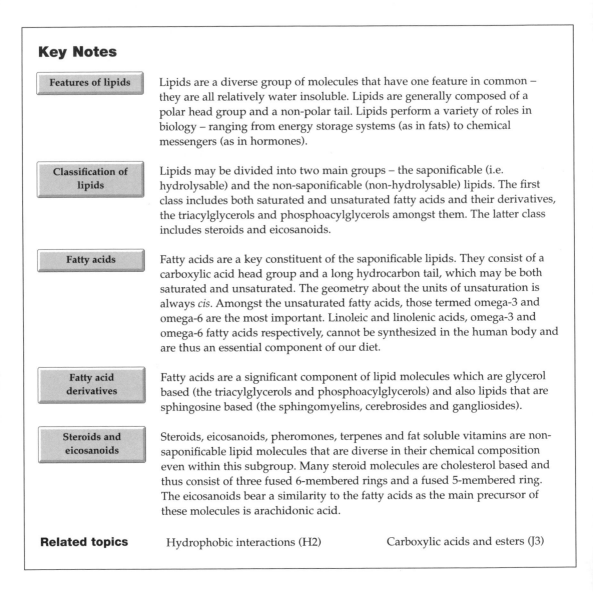

Key Notes

Features of lipids

Lipids are a diverse group of molecules that have one feature in common – they are all relatively water insoluble. Lipids are generally composed of a polar head group and a non-polar tail. Lipids perform a variety of roles in biology – ranging from energy storage systems (as in fats) to chemical messengers (as in hormones).

Classification of lipids

Lipids may be divided into two main groups – the saponifiable (i.e. hydrolysable) and the non-saponificable (non-hydrolysable) lipids. The first class includes both saturated and unsaturated fatty acids and their derivatives, the triacylglycerols and phosphoacylglycerols amongst them. The latter class includes steroids and eicosanoids.

Fatty acids

Fatty acids are a key constituent of the saponifiable lipids. They consist of a carboxylic acid head group and a long hydrocarbon tail, which may be both saturated and unsaturated. The geometry about the units of unsaturation is always *cis*. Amongst the unsaturated fatty acids, those termed omega-3 and omega-6 are the most important. Linoleic and linolenic acids, omega-3 and omega-6 fatty acids respectively, cannot be synthesized in the human body and are thus an essential component of our diet.

Fatty acid derivatives

Fatty acids are a significant component of lipid molecules which are glycerol based (the triacylglycerols and phosphoacylglycerols) and also lipids that are sphingosine based (the sphingomyelins, cerebrosides and gangliosides).

Steroids and eicosanoids

Steroids, eicosanoids, pheromones, terpenes and fat soluble vitamins are non-saponificable lipid molecules that are diverse in their chemical composition even within this subgroup. Many steroid molecules are cholesterol based and thus consist of three fused 6-membered rings and a fused 5-membered ring. The eicosanoids bear a similarity to the fatty acids as the main precursor of these molecules is arachidonic acid.

Related topics

Hydrophobic interactions (H2)   Carboxylic acids and esters (J3)

Features of lipids   The name 'lipid' comes from the Greek 'lipos' and literally means fat; however fats are just one subgroup of this very diverse class of molecules. Lipids are important biological systems that, although they differ in composition, have two inter-related features; they all contain a polar head group and a non-polar tail and they are all relatively water insoluble:

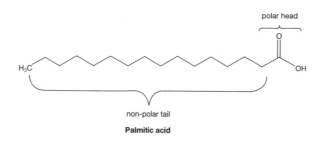

Palmitic acid

Lipids perform a variety of roles. They can act as storage systems of chemical energy that comes from the food we eat. Lipids are involved in maintaining our body temperature and in protecting important tissue and organs from mechanical damage. They are of course a major component of all cell membranes. One of the more specialized functions of lipids is to act as chemical messengers, as in hormones which are lipids.

Classification of lipids

Lipids vary greatly in their structure and composition, however they may be categorized as belonging to one of two major groups, based solely on their susceptibility to hydrolysis under alkaline conditions (see Topic J3). The two groups are termed **saponifiable** and **non-saponifiable** lipids. Saponification (a soap-forming reaction) refers to the hydrolysis process which converts an ester to the salt of a carboxylic acid, and which takes place in the presence of a base. The saponifiable lipids include saturated and unsaturated fatty acids and their derivatives (see *Fig. 1*). The non-saponifiable lipids include steroids and eicosanoids.

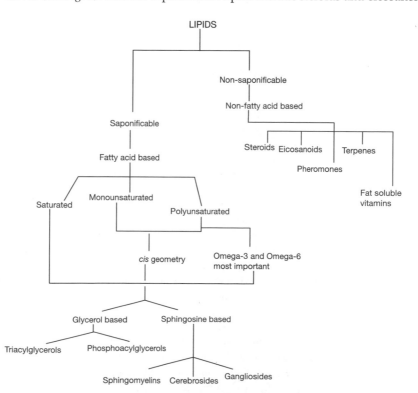

Fig. 1. The diversity of the molecules referred to as lipids.

Fatty acids

The term 'fatty acid' follows from the fact that all molecules of this type have a carboxylic acid head group and a long, unbranched hydrocarbon, wax-forming (and hence 'fatty') tail (see *Table 1*). The fatty acids that are synthesized in the body come from acetic acid (CH_3COOH) and consequently always involve an even number of carbons in their chain (including the carboxyl carbon). For example, palmitic acid is a C-16 fatty acid. The hydrocarbon tail may be saturated, as in palmitic acid, monounsaturated, as in oleic acid, or polyunsaturated, as in linoleic acid. Whatever the level of unsaturation, in naturally occurring and biologically important fatty acids, the geometry about the carbon–carbon double bond is always *cis* (see Topic E1, and E2).

Table 1.   Examples of fatty acids

Structure information	Common name	Structure
18:0	steric acid	
18:1	oleic acid	
18:2	linoleic acid	
18:3	linolenic acid	
20:4	arachidonic acid	

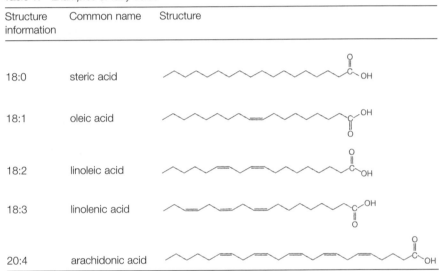

Amongst the polyunsaturated molecules the so-called omega-3 and omega-6 fatty acids are the most important. EPA (eicosapentaenoic acid) is an example of an omega-3 fatty acid and linoleic acid an example of an omega-6 system. The numbering of these molecules indicates the position of the last carbon–carbon double bond in the tail ('omega', as this is the last letter in the Greek alphabet). The number '3' indicates the last double bond is 3 carbons from the end of the tail; the '6', six carbons from the end. In addition the symbol delta (Δ) and numerical subscripts indicates the position of each of the double bonds in the chain (linoleic acid could be described as $18:2^{\Delta9,12}$, where '18' is the carbon chain length and '2' refers to the number of double bonds). Amongst the important polyunsaturates, two molecules are a vital component of our diets. Linoleic and linolenic acids cannot be synthesized in the human body and hence are required in our foodstuff. These fatty acids are therefore termed '**essential**'.

Fatty acid derivatives

Fatty acids are a significant component of a large number of saponifiable lipids. Fats, oils, phosphoacylglycerols, sphingolipids, sphingomyelins, cerebrosides and gangliosides, all contain a fatty acid moiety in their structure.

The organic-chemical class to which the fats and oils belong is that of esters, as they are formed from the reaction between a carboxylic acid (the fatty acid) and an alcohol (more specifically a triol called **glycerol**) (see Topic J3) (see *Fig. 2*).

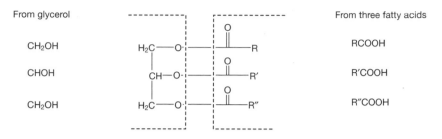

Fig. 2.   *The general structure of fats and oils.*

In nature it is more often found that the triacylglycerol is the result of a reaction between glycerol and more than one type of fatty acid, hence there is a wide variety of fats and oils. To differentiate between fats and oils it is simply necessary to consider their physical state. Fats are solid or semi-solid at room temperature as they are formed from triacylglycerols that contain a high percentage of saturated fatty acids that form long extended chains capable of the close packing associated with the solid state. Oil-forming triacylglycerols have a higher percentage of mono- and polyunsaturates that are more varied in their overall shape, hence not capable of close packing, resulting in a liquid at room temperature. The triacylglycerols are considered **metabolic fuel**.

Phosphoacylglycerols are also based on glycerol and fatty acids but with one of the fatty acids being replaced by phosphoric acid, or rather an ester of phosphoric acid (see *Fig. 3*).

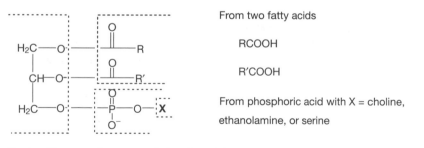

Fig. 3.   *The general structure of phosphoacylglycerols.*

The alcohol of the phosphoester may be choline (OH$\underline{C}$H$_2$CH$_2$N$^+$(CH$_3$)$_3$), ethanolamine (HO$\underline{C}$H$_2$CH$_2$NH$_2$) or serine (HO$\underline{C}$H$_2$CH(COOH)(NH$_2$)) ('C' in bold and underlined is the point of attachment to the phosphoric acid, with loss of water) although these are generally charged under physiological conditions. Although similar to the triacylglycerols, these phospho- compounds perform a very different role being the most abundant component of cell membranes. The ethanolamine and serine derivatives are also referred to as **cephalins** and are found in high concentrations in the brain.

The **sphingolipids** are a non-glycerol based subgroup of the saponificable lipids, they differ from each other in the nature of the groups attached to a **sphingosine** backbone:

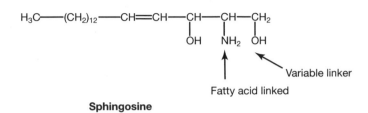

Sphingosine

All sphingolipids have a fatty acid attached via an amide link to the central amine and another group attached to the terminal –OH. They differ in the nature of this second group. If the terminal group is a phosphacholine, in which phosphoric acid is linked via ester formation to the sphingosine and choline attached to the phosphoric acid via another ester (as above) then a **sphingomyelin** is formed; these molecules are found in all cell membranes.

If the second group is a simple carbohydrate then these lipids are called **cerebrosides**; when more complicated carbohydrates are attached such molecules are referred to as **gangliosides**: (see *Fig. 4*).

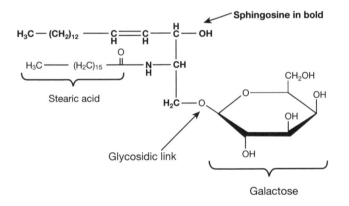

Fig. 4.   An example of a ganglioside.

Both of these classes of lipid are found in the grey matter of the brain.

Steroids and eicosanoids

Steroids and eicosanoids are two of a group of five non-saponificable lipids that we know about, the others being terpenes (responsible for many of the fragrances that are characteristic of fruits, for example), pheromones, and fat soluble vitamins (such as vitamin A or retinol). Amongst these only the eiscosanoids bears a resemblance to the lipids discussed thus far.

Steroids are molecules that are built of a series of fused 6-membered and 5-membered carbon rings. Cholesterol is probably the most familiar of them, it is the most abundant steroid in the body:

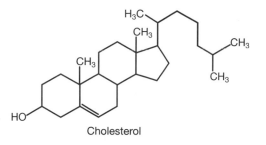

Cholesterol

Cholesterol is found in cell membranes and in blood and is vital for the synthesis of a number of hormones (see *Table 2*). Our obsession with cholesterol stems from the fact that the body makes enough of this steroid for its needs, so any cholesterol we take in our diet is surplus to requirements. It is now generally accepted that there is a link between high cholesterol levels and coronary heart disease.

Some hormones, the body's chemical messengers, are steroids based on the cholesterol framework. Steroid based hormones include the sex hormones; the estrogens, androgens and progestins, and the other regulatory hormones called the adrenocortical hormones.

The eicosanoids bear a similarity to the fatty acids and that is because the precursor for most of them is arachidonic acid (a C20 molecule). Their name comes from the Greek 'eikos' which means 20. Eicosanoids are found in all cells except blood cells and are implicated in a variety of processes from inflammation to the induction of labor! Two of the most important types of eicosanoid are the prostaglandins and the thromboxanes (see *Table 2*).

Table 2. Some examples of steroids and eicosanoids.

Common name	Structure

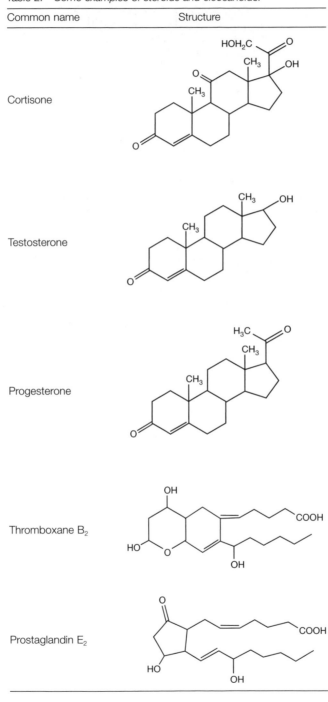

Cortisone

Testosterone

Progesterone

Thromboxane B_2

Prostaglandin E_2

L1 AROMATICITY

Key Notes

Benzene

Benzene is an unsaturated molecule and, as such, would be expected to undergo reactions similar to those of other unsaturated hydrocarbons such as alkenes and alkynes. However, benzene is relatively inert, and when it does react favors substitution reactions over addition reactions. The unexpected chemical and physical properties of benzene may be explained by the concept of pi electron delocalization. Benzene is the classic example of an aromatic compound. The term aromatic is applied as benzene, and other ring systems which have similar delocalized pi systems, is fragrant.

Molecular orbital description of benzene

Benzene is a planar molecule in which all of the bond angles about the carbon atoms are 120°. This bond angle is what would be expected for an sp^2 hybridized carbon atom, and therefore means that at each of the six carbon atoms there is a singly occupied p-orbital. These p-atomic orbitals overlap to form six pi molecular orbitals. The molecular orbital picture of benzene helps explain the special stability of this molecule.

Definition of aromaticity

In 1931 the physicist Erich Hückel carried out a series of calculations based on the molecular orbital picture of benzene, but extended this to cover all planar monocyclic compounds in which each atom had a p-orbital. The results of his work suggested that all such compounds containing (4n + 2) pi electrons should be stabilized through delocalization and therefore should also be termed aromatic.

Related topics   Factors affecting reactivity (I3)   Natural aromatics (L2)

Benzene

The study of the class of compounds now referred to as **aromatics** began in 1825 with the isolation of a compound, now called **benzene**, by Michael Faraday. At this time the molecular formula of benzene, C_6H_6, was thought quite unusual due to the low ratio of hydrogen to carbon atoms. Within a very short time the unusual properties of benzene and related compounds began to emerge. During this period, for a compound to be classified as aromatic it simply needed to have a low carbon to hydrogen ratio and to be **fragrant**; most of the early aromatic compounds were obtained from balsams, resins or essential oils. It was sometime later before Kekulé and coworkers recognized that these compounds all contained a six-carbon unit which remained unchanged during a range of chemical transformations. Benzene was eventually recognized as being the parent for this new class of compound. In 1865 Kekulé proposed a structure for benzene; a six-membered ring with three alternating double bonds (see *Fig. 1*). However, if such a structure were correct then the addition of two bromine atoms to adjacent carbons would result in the formation of two isomers of 1,2–dibromobenzene (see *Fig. 1*). Only one compound has ever been found. To account for this apparent anomaly Kekulé suggested that these isomers were in a state of rapid equilibrium

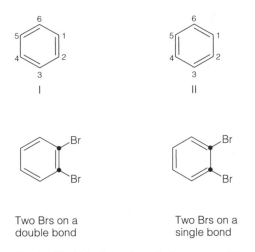

Two Brs on a   Two Brs on a
double bond   single bond

Fig. 1.   The Kekulé structures for benzene and the proposed isomers of 1,2-dibromobenzene.

(see *Fig. 1*). This is now known to be incorrect, and the Kekulé structures do not actually represent the true structure of benzene. Nevertheless, they are still routinely used, particularly when illustrating reaction mechanisms involving such compounds.

It was not until the **resonance** approach (see Topic I3) for looking at bonds was developed that a structure for benzene could be produced, from which its stability and chemical reactivity could be explained.

The basic postulate of resonance theory is that whenever two or more **Lewis structures** (see Topic B2) can be written for a molecule differing only in the positions of electrons, none of the structures are in complete agreement with the compounds chemical and physical properties. Thus, neither of the Kekulé forms of *Fig. 1* correctly reflect the properties of benzene, rather the true picture is somewhere in between. The 'structures' labeled **I** and **II** in *Fig. 1* (from now on called **canonical forms** as they are not isolatable molecules) when blended produce a **hybrid** structure (see *Fig. 2*), in which the carbon-carbon bonds are

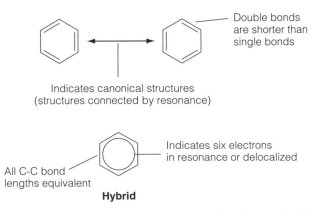

Double bonds
are shorter than
single bonds

Indicates canonical structures
(structures connected by resonance)

Indicates six electrons
in resonance or delocalized

All C-C bond
lengths equivalent

Hybrid

Fig. 2.   The Kekulé structures of benzene are actually canonical forms, the combination of which results in a hybrid structure which more accurately depicts the true structure of the molecule.

neither single nor double, but somewhere in between; this has now been determined experimentally. In this structure the pi electrons are not associated with particular atoms, rather they are **delocalized** over the six carbon centers making each carbon identical. The more resonance or canonical forms of **equivalent energy** that may be written for a molecule, the more stable the molecule. Therefore, the extra stability associated with benzene is called **resonance energy** (see *Fig. 3*).

Very early on it was determined that although benzene had several units of unsaturation, it did not display the same reactivity as alkenes or alkynes to oxidizing or reducing agents (see Topic I2), nor did it undergo the electrophilic addition reactions characteristic of such compounds (see Topics I1 and I2). This behavior could also be rationalized by resonance theory and the hybrid structure for benzene.

Molecular orbital description of benzene

Another approach to explain the structure of benzene and related compounds involves the use of **molecular orbitals** (MO's) (see Topic B2). The bond angles and bond lengths determined for benzene are all in agreement with the carbon centers being sp^2 hybridized. Therefore, each carbon atom has a singly occupied p-orbital associated with it. The dimensions of the molecule are such that the six p-orbitals are close enough to each other to overlap; as all of the carbon-carbon bond lengths are the same the p-orbitals overlap equally around the ring.

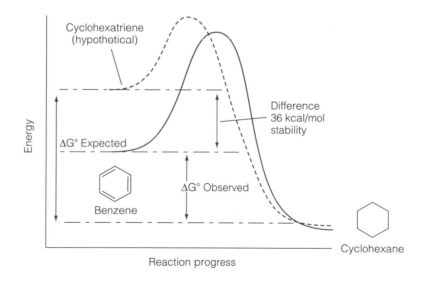

Heats of hydrogenation ($\Delta H°$ hydrog)

Reactant	Product	$\Delta H°$ hydrog (kcal mol^{-1})
Cyclohexene	Cyclohexane	28.6
1,3 Cyclohexadiene	Cyclohexane	55.4
Benzene	Cyclohexane	49.8

Fig. 3. The relative stabilities of cyclohexene, 1,3-cyclohexadiene, 1,3,5-cyclohexatriene, and benzene.

According to **molecular orbital theory** (see Topic B2), six overlapping p-orbitals will combine to form a set of six pi molecular orbitals (see *Fig. 4*). In benzene all of the bonding molecular orbitals are doubly occupied, and there are no electrons in the antibonding MO's. Benzene is thus said to have a closed shell of delocalized pi electrons, this accounts in part for the stability of benzene.

Definition of aromaticity

Before 1900, chemists assumed that any ring with alternating single and double bonds would display properties similar to the parent aromatic compound, benzene. However, the synthesis of cyclooctatetracene; an eight membered ring with four alternating single and double bonds, meant that this idea had to be discounted. Cyclooctatetracene displays the same chemistry as alkenes. It was not until the 1930's that aromaticity could be predicted.

In 1931 Erich Hückel performed a series of calculations based on the molecular orbital picture for benzene. In his original work he only considered monocyclic compounds, although later his ideas were extended to polycyclic systems. His results showed that a planar monocyclic system containing **(4n + 2)** pi electrons; where n = 0, 1, 2, 3, . . . , would have a closed shell of 2, 6, 10, 14, . . . , delocalized electrons as in benzene. This (4n + 2) requirement is now know as **Hückel's rule** for aromaticity. As cyclooctatetracene has 8 pi electrons Hückel would have predicted that this compound was nonaromatic. Hückel's rule can readily be

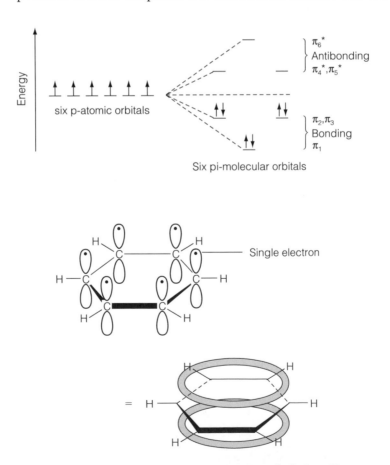

Fig. 4.   The molecular orbital analysis and pi electron distribution of benzene.

applied to cyclic compounds containing atoms other than simply carbon. For example, pyridine, pyrrole, furan, and thiophene, all have 6 delocalizable pi electrons and are therefore aromatic. These, together with a number of other aromatic systems (see *Fig. 5*), are the parent structures for a range of biologically or pharmaceutically relevant compounds.

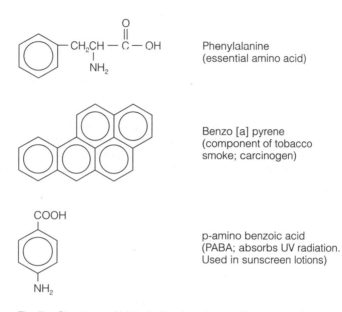

Phenylalanine
(essential amino acid)

Benzo [a] pyrene
(component of tobacco
smoke; carcinogen)

p-amino benzoic acid
(PABA; absorbs UV radiation.
Used in sunscreen lotions)

Fig. 5.   Structures of biologically relevant aromatic compounds.

L2 NATURAL AROMATICS

Key Notes

Aromatic amino acids
There are four naturally occuring aromatic α-amino acids; tyrosine, phenylalanine, histidine and tryptophan. Both tyrosine and phenylalanine are benzene-based 6-pi electron systems. Tryptophan is an indole-based system, and histidine is an imidazole-based system.

Nitrogen heteroaromatics in cofactors
Nitrogen based aromatics appear in many essential cofactors and coenzymes. The pyridine ring system is the basis for a number of the B vitamins, as is the pyrimidine and pteridine (bicyclic) system. A tricyclic derivative of the pteridine system, vitamin B_2, is a prosthetic group on a number of dehydrogenases.

Nucleic acids
The nucleic acids, DNA and RNA, are composed of heteroaromatic subunits. These subunits are purine and pyrimidine based. The interaction between the delocalized pi systems of adjacent residues is an important contributor to the overall stability of double helical DNA.

Aromaticity in biochemical reactions
A well characterized biochemical reaction involving an aromatic compound is the role of pyridoxal phosphate in the transamination of amino acids. One of the steps of this enzyme catalyzed reaction requires the nitrogen of the pyridine ring to be protonated, forming an intermediate stabilized by electron delocalization within the pi aromatic system. The ability of aromatic rings to delocalize charge in this way, is fundamental to their chemical and biochemical importance.

Related topics
Hydrophobic interactions (H2)
Factors affecting reactivity (I3)
Aromaticity (L1)

Oligonucleotide synthesis (M2)
Nucleic acids (K2)

Aromatic amino acids
Of the 20 naturally occuring α-amino acids four are aromatic (see Topic L1 and *Fig. 1*).

Phenylalanine and tyrosine are clearly both benzene based, with their aromaticity arising from the six pi electrons in the six carbon-centered ring. In addition to the aromatic character of tyrosine, its side chain can also act as a weak acid. When the 'OH' group is deprotonated, the resultant negative charge (pair of electrons) on the oxygen atom becomes part of the delocalized pi system and a number of **resonance** or **canonical** forms (see Topic I3) for this ion may be drawn (see *Fig. 2*).

The sidechain of tryptophan is an indole ring; that is a benzene ring fused to a 5-membered ring containing a nitrogen atom and a carbon-carbon 'double bond'. Eight of the electrons of the aromatic system are associated with carbon centers, the final two being the lone pair of electrons on the nitrogen atom. Consequently, although the indole sidechain has an amine component (see Topics J4 and M1), this amine cannot function as a nucleophile or a base (see Topic I1) as the lone pair of electrons is not available. In contrast, the imidazole ring of histidine (see Topic

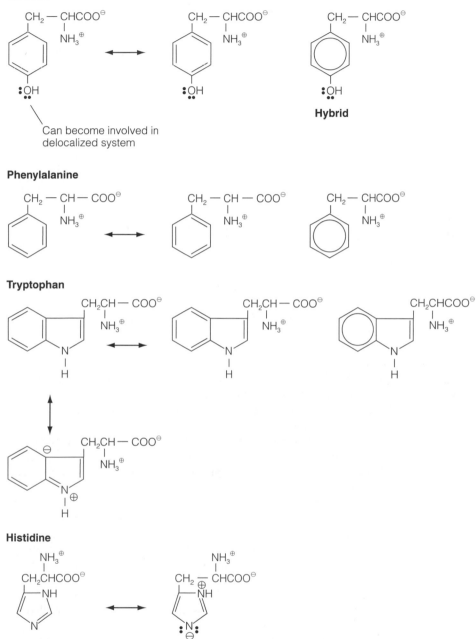

Fig. 1. *The aromatic amino acids; their canonical forms and resonance hybrids.*

M1) has two nitrogens but only one of these donates its lone pair for the 5-membered ring to attain aromaticity. Consequently histidine can act as a base.

It should be noted that humans cannot synthesize benzene rings, and consequently both phenylalanine and tryptophan are essential components of our diet. Phenylanlanine may be converted to tyrosine in a reaction catalyzed by the

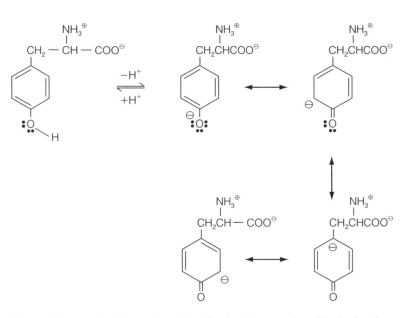

Fig. 2.   The canonical forms of tyrosine following deprotonation of the hydroxyl group.

enzyme phenylalanine hydroxylase, and therefore tyrosine need not be ingested directly.

Nitrogen heteroaromatics in cofactors

Nitrogen based aromatics display a huge range of functions in biological systems (see *Fig. 3*). In particular the pyridine ring is the basis for nicotinamide, the amide of nicotinic acid. This and the purine derivative, adenine, are present in one of the most important coenzymes in biological oxidations; nicotinamide adenine dinucleotide (NAD^+).

Another pyridine based system is pyridoxal phosphate, a cofactor required for the decarboxylation and transamination of amino acids.

More complicated nitrogen aromatics also have a vital role. For example, the pterdine ring system (see *Fig. 3*) is found in vitamin B_{12}, folic acid. This vitamin is added to many food stuffs and is prescribed to all women, in advance and during the first 3 months of pregnancy, to reduce the possibility of neural tube defects in the unborn child.

A tricyclic nitrogen aromatic, Riboflavin (vitamin B_2) occurs as the prosthetic group in a number of dehydrogenases; enzymes which oxidize by removing hydrogen. In many of these systems the role of the aromatic ring is to stabilize, by delocalization, the charge (an excess or deficiency of electrons) which builds up during the course of the chemical transformation of a group connected to the ring.

Nucleic acids

Nucleotides (see Topic M2) are the building blocks of the nucleic acids, the molecules of hereditary. The aromatic component of nucleotides are the purine and pyrimidine ring systems. The purine rings of both adenine and guanine are 10 pi electron systems, with nitrogen lone pairs still available for these to act as amine bases. The pyrimidines of cytosine, uracil and thymine are all 6 pi electron systems, again with nitrogen lone pairs available for base acitivity (see *Fig. 4*). The delocalized system is obvious in the first set of structures of *Fig. 4*, however the more usual way to represent these molecules is as the corresponding **tautomer** (see *Fig. 4* and Topic J2).

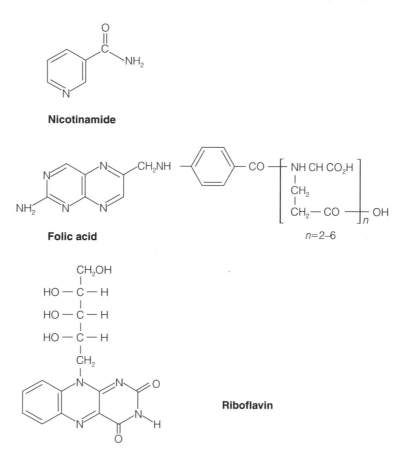

Nicotinamide

Folic acid

n=2–6

Riboflavin

Fig. 3.   Nitrogen-based aromatics in cofactors and coenzymes.

When nucleotides are incorporated into chains to form double helical DNA, the double helix is not only stabilized by hydrogen bonding between complementary heterocyclic bases (see Topic H1), but also by the favorable interactions between the pi electrons above and below the plane of successive aromatic rings (see Topic H2); pi-stacking is said to occur.

Aromaticity in biochemical reactions

The role of pyridoxal phosphate in the enzyme catalyzed transamination of amino acids is now well understood. The first step of this reaction (see *Fig. 5*) involves the condensation of the aldehyde on the pyridoxal ring with the amino group of the amino acid to form an imine (see Topic J2). An acid catalyzed **tautomerization** followed by hydrolysis, leads to the release of an α-keto (or oxo) acid and the formation of pyridoxamine phosphate. The pyridoxamine can then go on to react with a different keto-acid *via* the reverse reaction, and by transfering the amine group create another amino acid.

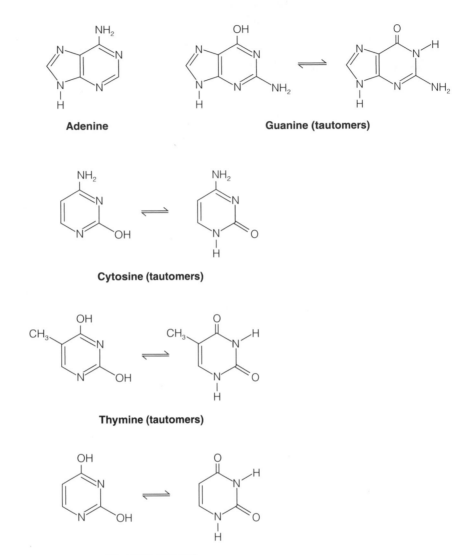

Fig. 4.   The heterocyclic bases of the nucleic acids including their tautomers.

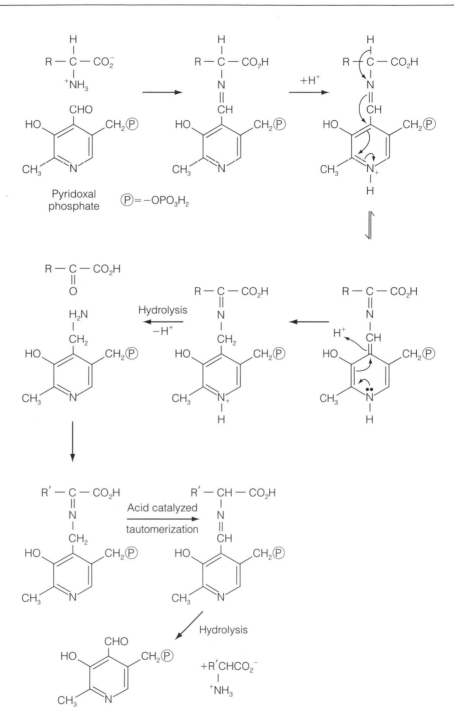

Fig. 5. The mechanism of amino acid transamination via pyridoxal phosphate.

M1 PEPTIDE SYNTHESIS

Key Notes

Natural amino acids

Compounds referred to as α-amino acids consist of an amino group and a carboxylic acid group, attached to the same carbon. There are 20 naturally occuring amino acids which all have an α-hydrogen, and differ only in the nature of the fourth substituent; referred to as the sidechain (R). Nineteen of the 20 naturally occuring amino acids are chiral and all of these have the L-configuration.

Functional group protection

In synthesizing peptides it is necessary to perform reactions between functional groups in a controlled/directed manner. It is therefore important that amino or carboxylic acid groups, and other types of functional groups present in sidechains, are protected, that is made nonfunctional, until required.

Coupling reactions

The reaction between an amino group and a carboxylic acid group to form an amide or peptide bond is not particularly rapid or efficient. To increase the efficiency of such a reaction additional reagents are involved; referred to as coupling reagents. These generally function by initially binding to the carboxylic acid group and in doing so making the carbonyl of this group more susceptible to the nucleophilic attack of an amine.

Solid phase synthesis

The chemical synthesis of all but the smallest peptides is both time and materials consuming as the growing peptide must be purified after each coupling step. A major step forward in peptide chemical synthesis came with the development of the Merrifield solid-phase technique. This method involves the use of an activated polymer to which the 'C-terminal' amino acid is attached, with its amino group available for amide coupling reactions. As the growing peptide remains attached to the polymer until the last coupling is complete, only one purification step is involved thus speeding up the whole synthetic process.

Related topics

Carboxylic acids and esters (J3)
Proteins (K1)

Amines and amides (J4)

Natural amino acids

The building blocks of proteins, the key functional molecules in biological systems, are the **α-amino acids**. There are 20 naturally occuring α-amino acids, all of which have the general structure shown in *Fig. 1*. These molecules differ only in the nature of the sidechain (R) (see *Table 1*). All but one of the amino acids have a chiral α-carbon (i.e. a carbon with four different groups attached) with the configuration of these compounds being uniformly L (see Topic E2)

As illustrated in *Table 1*, in many instances the sidechains of these molecules have additional functional groups. On the basis of these, the amino acids are generally subdivided into those that have basic, acidic, and aromatic characteristics. In combining two amino acids to form a dipeptide, an amino group reacts with a

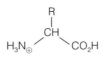

Fig. 1. *General structure of α-amino acids.*

Table 1. *The 20 most common amino acids; their structures, names, abbreviations, and sidechain properties*

	Name	Abbreviations	Property
— R	Name	Abbreviations	Property
— CH_3	Alanine	Ala, A	Hydrophobic
—$(CH_2)_3NHC$ $\overset{NH}{\underset{NH_2}{}}$	Arginine	Arg, R	Hydrophilic
— CH_2OCNH_2	Asparagine	Asn, N	Hydrophilic
— CH_2—CO_2H	Aspartic acid	Asp, D	Hydrophilic
— CH_2SH	Cysteine	Cys, C	Hydrophilic
— $CH_2CH_2CH_2CH_2NH_2$	Lysine	Lys, K	Hydrophilic
— $(CH_2)_2$—CO_2H	Glutamic acid	Glu, E	Hydrophilic
— $(CH_2)_2$—$CONH_2$	Glutamine	Gln, Q	Hydrophilic
— H	Glycine	Gly, G	Hydrophobic
—CH_2 (imidazole ring)	Histidine	His, H	Aromatic
— $CH(CH_3)CH_2CH_3$	Isoleucine	Ile, I	Hydrophobic
— $CH_2CH(CH_3)_2$	Leucine	Leu, L	Hydrophobic
— $CH(CH_3)_2$	Valine	Val, V	Hydrophobic
— $CH_2CH_2SCH_3$	Methionine	Met, M	Hydrophilic
— CH_2OH	Serine	Ser, S	Hydrophilic
— $CH(OH)CH_3$	Threonine	Thr, T	Hydrophilic
—CH_2 (phenyl ring)	Phenylalanine	Phe, F	Aromatic
—CH_2 (phenol ring)—OH	Tyrosine	Tyr, Y	Aromatic
—CH_2 (indole ring)	Tryptophan	Trp, W	Aromatic

carboxyl group with the result that water is lost; this type of reaction is known as a **condensation reaction** (see *Fig. 2*).

The reactions of amines and carboxylic acid groups are dealt with in detail in Topics J3 and J4. It should be remembered, that a number of sidechains also have carboxylic acid or amine functions that may undergo reactions identical to those of the amine and carboxylic acid group, required to form the growing peptide chain.

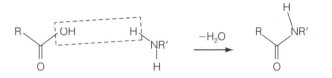

Fig. 2.   The formation of an amide or peptide bond results in the loss of a water molecule.

In addition, the amino acid cysteine has a free thiol (-SH) group in its side chain. This is readily oxidized to the disulfide, so care must be taken in handling this residue (see *Fig. 3*).

Amino acids can clearly display both acid and alkali properties. At neutral pH the carboxyl group is almost completely ionized (COO^-) as is the amino group (NH_3^+). This positive and negatively charged system is referred to as a **Zwitterion**. As amino acids are ionized under neutral conditions they have physical properties similiar to ionic compounds; that is, they are generally high melting point solids.

Functional group protection

In synthesizing a peptide or protein (either chemically or biochemically) it is important that the synthesis takes place in the correct direction. To illustrate, there are clearly three ways in which a glycine residue may be added to a lysine residue; an amide bond may be formed between the carboxyl of the glycine and the (a) α-amino group or the (b) ε-amino group of lysine, or between the (c) carboxyl of lysine and the glycine amino group. If the desired peptide was specified as Gly-Lys, then only the first of these would produce the desired product. Glycine being the first residue would have a free 'N-terminal', and lysine the last would have a free 'C-terminal', where 'N' and 'C' refer to α-amino and carboxyl groups. Side reactions, like (b) and (c) may only be avoided by capping or **protecting** groups not to be used in the required amide forming reaction.

Clearly amino, carboxyl, and other functional groups need to be protected, and perhaps more importantly subsequently deprotected in a controllable manner. One of the most efficient ways of protecting an amino group is to convert it to an amide! This is not useful in peptide synthesis, as in removing the protecting group

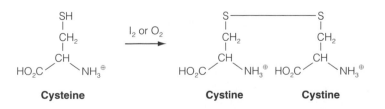

Fig. 3.   Cysteine residues are readily oxidized (i.e. hydrogen removed) to their disulfide equivalent. In protein sequences S-S bridged cysteines are referred to as cystines to differentiate between them and residues with free thiol (-SH) groups.

the peptide chain would be completely degraded. A good compromise is offered by the formation of alkoxyamides (see *Fig. 4*). These may be removed under conditions which are generally too mild to result in amide cleavage.

Protection of the carboxyl group is routinely achieved by converting it to the ester (see *Fig. 4*), which may also be removed under conditions too mild to cleave an amide bond. When a longer peptide chain is being prepared it is sensible to **selectively** remove protecting groups, clearly this requires the use of a range of protecting groups which are susceptible to cleavage under differing conditions. The use of complementary protecting groups is called **orthogonal protection**. In *Table 2* several examples of protecting group and cleavage conditions are provided.

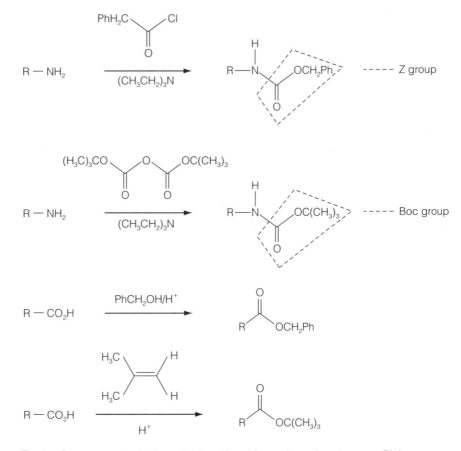

Fig. 4. *Common routes to the protection of α-amino and α-carboxyl groups. Ph is a benzene ring, (CH₃CH₂)₃ N is triethylamine (a base).*

Coupling reactions

The reaction of a carboxyl group with an amino group to form an amide bond results in the loss of water, the 'OH' of which comes from the carboxylic acid (see *Fig. 2*). The 'OH' group is not a good leaving group (see Topics I2, J1 and J3) and hence the carboxyl group must be **activated** for this reaction to proceed. One of the common approaches to activating such groups is to convert them to the acid chloride; this may be achieved by the addition of thionylchloride (SOCl₂) (see Topic I2). However, this is not the best approach for peptide synthesis as

Table 2.   Examples of protecting groups routinely used in peptide synthesis and the reagents required for their removal

Group	Amino acid[a]	Removal method
— NO₂	Arg— N	H₂/Pd— Charcoal
— CH₂Ph	Cys— S	Na/NH₃(liq)
— CH₂OCH₂Ph	His— N	H₂/Pd— Charcoal
— CH₂Ph	Ser— O	HBr/CH₃COOH or Na/NH₃(liq)
— CH₂Ph	Tyr— O	HBr/CH₃COOH or Na/NH₃(liq)

[a]S, N, O – refer to atoms in amino acid sidechains to which protecting
group is attached.

unwanted side reactions often occur, amino acid **racemization** being the most
serious of these (see Topics E2 and E3) (see *Fig. 5*).

A number of precautions may be taken to help reduce the degree of racemization whilst still promoting the coupling process. Reactions performed at low

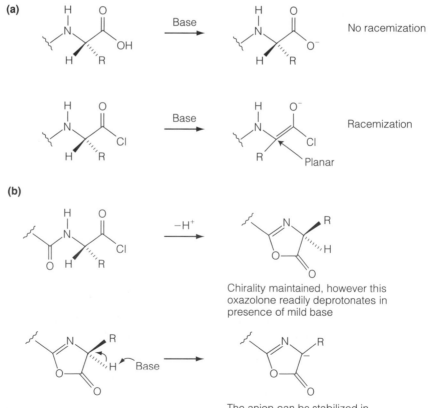

*Fig. 5.   An illustration of how racemization may occur in (a) activated carbonyl compounds in
general, and (b) peptides in particular.*

temperature, in solvents of low polarity and at neutral pH involve a reduced risk of racemization. Also it is has been found that if the amino group (alpha to the reacting carbonyl) has been alkoxycarbonyl protected, racemization is almost reduced to zero. There are a number of alternatives to acid chloride formation, all of which are in common use in peptide synthesis. These involve the use of coupling agents such as **dicyclohexylcarbodimide** (DCCI, also abbreviated as DCCD), and the preparation of anhydrides, mixed anhydrides, active esters and azides (see *Table 3*).

Table 3.　The main methods of peptide coupling; the advantages and disadvantages of each approach

Method	Activated form	Advantages	Disadvantages
DCCI		Simple to attach	Side reactions possible
Azide		No racemization	Side reactions possible
Anhydride		Very clean	Uses large amount of reagents therefore expensive
Active ester		Very clean	Expensive and/or difficult to prepare

Solid phase synthesis

The chemical synthesis of short peptides may be readily performed by simply mixing reagents in a flask. The reactions are generally efficient and any unreacted material or failure sequences (peptides which do not have the desired sequence), may be removed by standard chemical purification procedures. When longer sequences are required, this process becomes very time- and material-consuming; the growing peptide needs to be isolated and purified after each coupling step. However, the process can be greatly simplified using the **solid-phase synthesis** method introduced by R. B. Merrifield in the 1960s. The '**solid-phase**' referred to is a polymer of polystyrene beads, prepared so that one of every 100 or so benzene rings bears a chloromethyl ($-CH_2Cl$) group. This group is covalently linked to the carboxyl group of an amino acid, and then subsequent amino acids are added in a stepwise fashion to form a peptide chain attached to the polystyrene bead (see *Fig. 6*). The stepwise nature of this method means that it has leant itself to automation, and computer controlled peptide synthesizers are now in routine use. Each step of the process occurs with extremely high yield, and as the growing chain need only be purified following the final coupling, very little material is lost during sample handling.

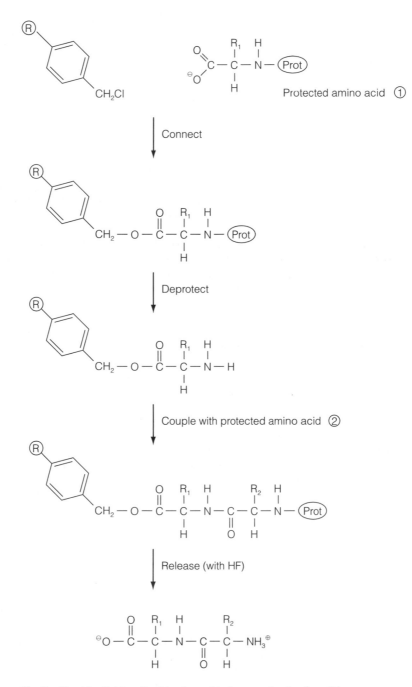

Fig. 6.   The Merrifield method for the solid-phase synthesis of peptides.

M2 OLIGONUCLEOTIDE SYNTHESIS

Key Notes

Heterocyclic bases	Heterocyclic compounds are compounds involving rings, with a least one member of the ring being an atom other than carbon. The most important of such compounds, from a biological viewpoint, are the pyrimidine and purine systems. As each of these contain nitrogen atoms with lone pairs of electrons, they are often referred to as heterocyclic bases.
Nucleosides and nucleotides	When purine or pyrimidine rings are covalently linked to a furanose ring, a nucleoside is the result. When this molecule is further substituted with a phosphate group at the 5′ position, a nucleotide is the result. Nucleotides are the building blocks of the nucleic acids DNA (deoxyribonucleic acid) and RNA (ribonucleic acid).
Functional group protection	Nucleotides have many functional groups in addition to those required for their linking to form oligonucleotides. To avoid side reactions during the chemical synthesis of DNA or RNA molecules these functional groups must be protected, and the protecting groups be removable under conditions that will not degrade the growing nucleic acid chain. There are several ways in which this may be achieved and the chemical synthesis of small deoxyribo- and ribonucleic acid molecules is now routine.
Automated synthesis of nucleic acids	The development of protecting group chemistry and the chemistry required to promote the linkage between nucleotides, has been such that the whole process may be automated. This is similar to the Merrifield approach for peptide synthesis, and involves the use of a solid phase support for the growing oligonucleotide chain.
Related topics	Phosphoric acid and phosphates (F1) Natural aromatics (L2) Nucleic acids (K1)

Heterocyclic bases

Ring systems involving one or more atoms other than carbon are referred to as heterocyclics. Two heterocyclic systems are of fundamental importance in biology; these are the aromatic (see Section L) pyrimidine and purine rings (see *Fig. 1*). The most commonly occuring of these are adenine, guanine and cytosine in both deoxyribonucleic (DNA) and ribonucleic (RNA) acids, and thymine in DNA and uracil in RNA.

In each of these ring systems the nitrogen atoms have lone pairs of electrons associated with them, making these systems nucleophiles and bases, hence they are commonly referred to as the heterocyclic bases.

Nucleosides and nucleotides

When a heterocyclic base is covalently linked to a cyclic sugar molecule a nucleoside is said to be formed (see *Fig. 2*). There are two types of sugar involved in naturally occuring nucleosides. Both of these are ribose based and differ only in

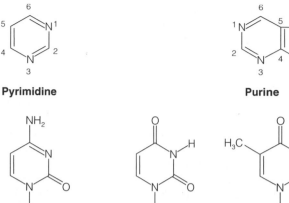

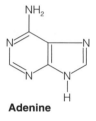

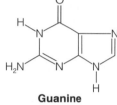

Pyrimidine   Purine

Cytosine   Uracil (RNA)   Thymine (DNA)

Adenine   Guanine

Fig. 1.   The pyrimidine and purine ring systems.

2′-deoxyribonucleoside   Ribonucleoside

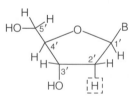

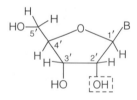

B=adenine, guanine   B=adenine, guanine
cytosine, thymine   cytosine, uracil

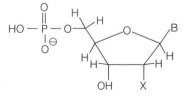

X=H, B=adenine, guanine   } 2′deoxyribonucleotide
cytosine, thymine

X=OH, B=adenine, guanine } ribonucleotide
cytosine, uridine

Fig. 2.   Nucleosides, and the nucleotides most commonly found in nucleic acids.

the presence or absence of an 'OH' group at the 2-position of the sugar ring. In deoxyribonucleic acids, the 'OH' is missing; hence 'deoxy', in ribonucleic acids the 'OH' is present. Once a bond is made between the base and the sugar, the carbon centers on the sugar are labeled 1', 2', etc. to distinguish between them and positions on the base. The sugar-base bond is made between the 1' C of the sugar and the N3 of pyrimidines, or N9 or purines, using the numbering shown in *Fig. 1*.

The addition of a phosphate group at the 5' position results in the formation of a nucleotide, the fundamental building blocks of DNA and RNA. To differentiate between nucleosides and nucleotides, when a phosphate group is attached to adenine it is referred to as adenosine. Likewise, guanine as guanosine, uracil as uridine, thymine as thymidine, and cytosine as cytidine (see *Fig. 2*).

In forming DNA or RNA molecules, or oligonucleotides in general, nucleotides are combined by a covalent bond between the 3' OH of the first nucleotide and the 5' phosphate of the second nucleotide. Consequently, a growing oligomer will have a free 5' phosphate on the first residue and a free 3' OH on the last residue. Oligonucleotides are therefore synthesized in the 5' to 3' direction (see *Fig. 3*).

Functional group protection

To develop an efficient route to the chemical synthesis of oligonucleotides, it has been necessary to establish protecting groups for the various functionalities

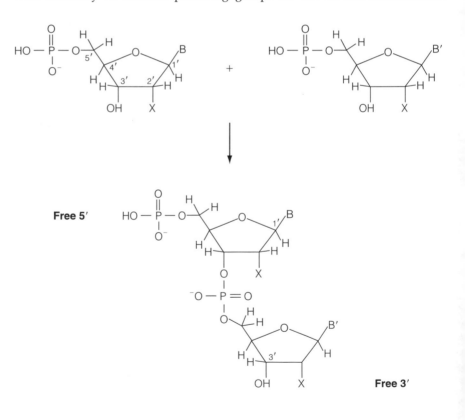

Fig. 3.   *A bond between the 3' OH and 5' phosphate results in the formation of a dinucleotide.*

present on both the heterocyclic bases and the ribose ring. It has also been necessary to devise groups which may be used to promote the coupling between nucleotides.

A number of synthetic routes are now in routine use, but perhaps the most common protecting groups utilized are those shown in *Fig. 4*. The amine function of adenine and cytidine is protected as a benzoyl (bz), that of guanine as an isobutyrl (ibu) moiety. Neither thymidine nor uridine need protection. When oligoribonucleotides are to be prepared it is necessary to protect the 2'-OH, and this is generally achieved by the addition of a tertiary butyl dimethylsilyl (TBDMS) group.

There is considerable diversity in the groups used to direct the 3' and 5' positions for subsequent coupling. However, one of the most common methods used

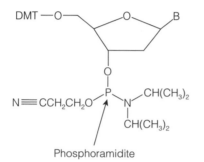

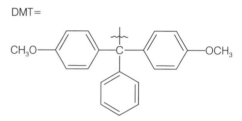

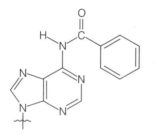

Phosphoramidite

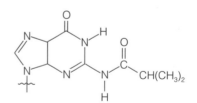

N-protected adenine

N-protected guanine

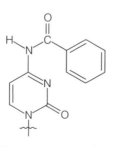

N-protected cytosine

Thymine does not require protection

Fig. 4.   The protected bases and activated sugars used in the phosphoramidite route to oligonucleotides.

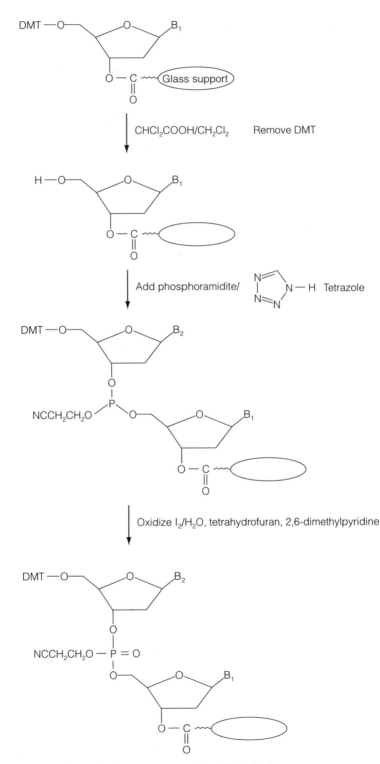

Fig. 4. continued

in the automated synthesis of oligonucleotides requires that the 5' position (with phosphate off) is activated (and protected) by the addition of a dimethoxytrityl (DMT) group, and the 3' position by the addition of a trivalent phosphorous group, to which good leaving groups are attached. Thus, this approach is referred to as the phosphoramidite method of oligonucleotide synthesis. This is currently the most commonly used and commercially developed, but not the only method in existence.

Automated synthesis of nucleic acids

The automated synthesis of deoxyribo- and more recently ribonucleic acids is now routine. Oligonucleotide synthesizers operate on a principle similar to that of the Merrifield solid-phase peptide synthesizer. A protected nucleotide is covalently linked to a solid phase support which is glass based. It is now usual to purchase the polymer support with the first (actually last in the desired sequence) protected nucleotide already attached. The synthesis proceeds with the removal of the 5' DMT group from this nucleotide and the addition of the next nucleotide prepared as the phosphoramidite. The sequence of events to follow are illustrated in *Fig. 5.*

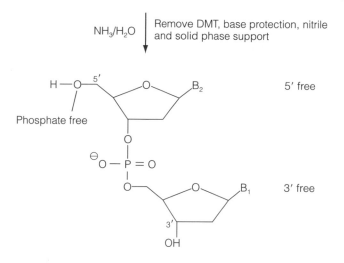

Fig. 5. Schematic representation of the solid-phase synthesis of oligonucleotides.

N1 LOWRY–BRONSTEAD ACID AND BASE

Key Notes

Definition

In the Lowry–Bronstead theory of acids and bases, an acid is defined as a substance which will give up a hydrogen ion (a proton) and a base is a substance which will accept a proton. This theory superceded that devised by Arrhenius and enabled the concept of acids and bases to extend beyond aqueous solutions.

Conjugate acids and bases

It is unlikely that protons can exist freely in solution. Consequently if an acid gives up a proton, there must be a base present to accept it. Thus, when an acid and base differ by one proton they are said to be conjugate to each other. Every acid must have its conjugate base and every base its conjugate acid.

Ionic strength

Dissolved ions, such as those produced by acid-base reactions, cannot be thought of as isolated entities. Rather, the properties of an ion are influenced by the presence of neighboring ions, owing to electrostatic forces between them. Consequently, it follows that the concentration of an ion in solution is not a true reflection of its ability to determine any property of the solution, except at infinite dilution when electrostatic interactions are minimized. To overcome this problem Debye and Hückel defined the term ionic strength, which depends not only on the concentration of all of the ionic species present, but also on the charge carried by each ion.

Related topic

Water – the biological solvent (C1)

Definition

The terms acid and base were originally defined by Arrhenius as substances which gave rise to hydrogen ions (H^+) and hydroxide ions (^-OH), respectively in solution. The combination of an acid and base would lead to neutralization and the formation of neutral water.

$$H^+ + {}^-OH \longrightarrow H_2O$$

Of course these definitions were only applicable to aqueous solutions. A more general theory was subsequently developed, by Lowry and Bronstead independently. The **Lowry–Bronstead** theory of acids and bases encompassed the definitions of Arrhenius and extended them to cover nonaqueous media.

According to the Lowry–Bronstead theory, an acid is any substance which can donate a proton and a base any substance which can accept a proton.

$$\underset{\text{acid}}{HA} \rightleftharpoons \underset{}{H^+} + \underset{\text{base}}{A^-}$$

Notice that an equilibrium process is indicated. This is because the propensity for an acid to give up a proton will be affected by the affinity of the base for that proton.

Conjugate acids and bases

Protons, and ions in general, are unlikely to be free in solution. Consequently, when an acid gives up a proton, it can only do so when there is something to accept it. In this way therefore one of the products of a reaction involving an acid (A_1) must be a **conjugate base** (B_1), and vice versa, reactions involving a base (B_2) must produce a **conjugate acid** (A_2).

$$A_1 + B_2 \rightleftharpoons A_2 + B_1$$

Some examples of acid–base reactions in which the acid or base is actually the solvent are provided in *Fig. 1*.

It is clear from the examples provided, that water can act as either an acid or a base depending on other species present (see Topic B3). These examples also illustrate the importance of the nature of the solvent in determining whether a substance will exhibit acid or base properties. With this in mind, solvents may be divided into three main classes;

(a) Protogenic (or protic) solvents — these are solvents which can give up protons and may behave as acids.
(b) Protophilic solvents — these are solvents which can accept a proton and thus act as bases.
(c) Aprotic solvents — these are solvents which can neither accept nor donate protons and thus prevent any solute displaying acid or base properties.

Ionic strength

Ions in solution cannot be thought of as isolated charged systems, rather their properties, and their effect on the solution, are influenced by the presence of neighboring ions with which there are electrostatic interactions. As a consequence

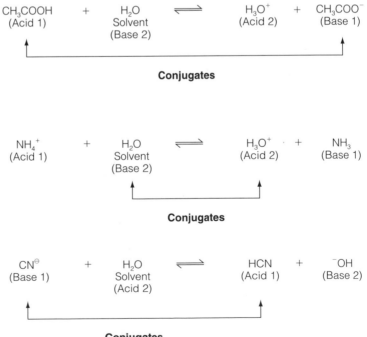

Fig. 1. Examples illustrating conjugate acid–conjugate base relationships and the role of the solvent.

of this, simply knowing the concentration of a particular ion (for the present purposes a proton (H^+, more accurately H_3O^+, the solvated proton)) does not adequately convey the consequences of this for the properties of the overall solution. This problem has led to the concept of **relative activity**, which takes account of both the concentration of the ion and interactions with its surroundings.

$$a = yc$$

a = relative activity
y = activity coefficient
c = concentration

Of course at infinitely low ion concentrations, electrostatic forces are reduced to a minimum.

The activity coefficient has been found to vary with ion concentration, in a rather complicated fashion. However, Debye and Hückel were able to derive an expression to describe this dependence;

$$\log y_i = -Az_i^2 \sqrt{I}$$

y_i = activity coefficient of ion type i
A = constant
z_i = charge on i
I = ionic strength of solution

The **ionic strength** of the solution is given by

$$I = \tfrac{1}{2} \Sigma c_i z_i^2$$

c_i = concentration of ion i

The summation is carried out over all types of ions present.

The constant A depends upon both the temperature and dielectric constant of the solvent.

In practice, the concentrations of ions in solution (for biological studies) is considered low and therefore the activity coefficient tends to unity, and thus concentrations may be used directly to reflect the properties of solutions.

N2 DISSOCIATION CONSTANTS OF ACIDS AND BASES

Key Notes

K_a	When an acid is dissolved in water it deprotonates to form its conjugate base. The process involved is in equilibrium, hence there are constants for the forward and backward reaction. K_a is the thermodynamic constant describing the dissociation of the acid and depends on the concentration of the proton (or solvated proton), the conjugate base, and inversely on the concentration of the acid.
K_b	By analogy to the dissociation constant for the ionization of an acid, K_b is the equivalent for base ionization (the reverse reaction in fact). Such that K_b (the thermodynamic constant) is dependent on the concentration of the protonated base (that is the conjugate acid of the base) and the base product (conjugate to the original proton source) and inversely proportional to the concentration of the original base.
Dissociation in terms of conjugate acids and bases	When dealing with acids and bases in aqueous solvents, it becomes clear that the product of the dissociation constants K_a and K_b, is equal to the ionic product of water, symbolized as K_w. This implies that in the same solvent, acids and conjugate bases have an inverse ratio of strengths. Consequently, it is the custom to simply refer to acid dissociation constants.
Related topics	Water – the biological solvent (C1) Lowry–Bronstead acid and base (N1)

K_a

When an acid (HA) dissolves in water an equilibrium is established between the acid, water acting as a base, and their respective conjugates. Any equilibrium has a thermodynamic constant associated with it and that for acid-base interconversions may be defined as follows;

$$HA + H_2O \rightleftharpoons H_3O^+ + A^-$$

$$\frac{a_{H^+} \times a_{A^-}}{a_{HA} \times a_{H_2O}} = constant$$

H^+ represents the hydrated proton, H_3O^+ and a is the relative activity for each of the species involved in the equilibrium process. It is more usual however, to assume that dilute solutions are involved and thus the concentration of the individual entities may be used rather than their activities (see Topic N1). Thus, an approximate value of the equilibrium constant for the forward reaction, K_a (the thermodynamic constant) may be defined as follows;

$$\frac{[H^+][A^-]}{[HA]} \approx K_a$$

[] indicate the concentration of the species enclosed.

As K_a values are very small numbers, a logarithmic scale is often referred to; the negative logarithm, to the base 10, of K_a is expressed as pK_a.

K_b

An analogous situation to that shown above for acids occurs when a base is dissolved in water; in this situation water will act as an acid (see Topic N1). An equilibrium is set up with a rate constant for the forward reaction which depends on the concentration of the protonated base, hydroxide ions, and is inversely dependent on the concentration of the original base (again with the assumption that dilute solutions are being employed).

$$B + H_2O \rightleftharpoons BH^+ + \ ^-OH$$
$$\frac{[BH^+][^-OH]}{[B]} \simeq K_b$$

Acid and base dissociation constants are useful in indicating their strength. The higher the dissociation constant the more significant the forward reaction; the greater the strength of the base in the equilibrium above.

As K_b values are very small numbers, a logarithmic scale is often referred to; the negative logarithm, to the base 10, of K_b is expressed as pK_b.

For information the dissociation constants for selected acids and bases are provided in *Table 1*.

Table 1. Dissociation constants for selected acids and bases in water at 25°C

Acid	K_a	Base	K_b
Acetic (CH_3COOH)	1.75×10^{-5}	Ammonia (NH_3)	1.77×10^{-5}
Formic ($CHOOH$)	1.77×10^{-4}	Ethylamine ($CH_3CH_2NH_2$)	4.66×10^{-4}
Benzoic (C_6H_5COOH)	6.3×10^{-5}	Diethylamine ($CH_3CH_2)_2NH$	1.00×10^{-2}
		Aniline ($C_6H_5NH_2$)	3.8×10^{-10}

Dissociation in terms of conjugate acids and bases

It has been noted that even very pure solvents display a very small electrical conductance, which can be rationalized by their tendency to produce ions by **self-ionization**. In the general case this may be represented as follows;

$$XH + XH \rightleftharpoons XH_2^+ + X^-$$

As the extent of ionization is very small, the activity of the un-ionized solvent is correspondingly small and hence the following expression may be derived;

$$[XH_2^+][X^-] \simeq K_I$$

K_I is known as the **ionic product**. Hence, for water the ionic product is given as follows;

$$H_2O + H_2O \rightleftharpoons H_3O^+ + \ ^-OH$$
$$[H_3O^+][^-OH] = K_w = [H^+][^-OH]$$

and K_w is the ionic product specifically for water. K_w for pure water at 25°C is 10^{-14}.

If the expression for the rate constants for acid and base dissociations are considered and their **product** derived, it can be seen that this is equivalent to the **ionic product** of water.

$$K_a = \frac{[H^+][A^-]}{[HA]} \quad K_b = \frac{[HA^-][^-OH]}{[A^-]}$$
$$K_a K_b = [H^+][^-OH] = K_w$$

This result suggests that the strengths of a conjugate acid and base in the same solvent must be in inverse ratio to one another; the greater the strength of the acid the weaker the strength of the conjugate base. Consequently, it is generally sufficient to talk about only one dissociation constant and this is usually that for acid ionization.

N3 ACIDITY AND ALKALINITY OF AQUEOUS SOLUTIONS

Key Notes

pH	In a sample of pure water, (solvated) hydrogen ions and hydroxide ions form by the process of self-ionization. Water is said to be neutral, as the concentration of each of these ions must be equivalent. Thus, as the ionic product for water, K_w, is 10^{-14} M at room temperature, the concentration of each of the ions must be 10^{-7} M under the same conditions. Hence, in general if the concentration of hydrogen ions in solution is greater than 10^{-7} M then the solution is said to be acidic, if the concentration of H^+ is less, then the solution is termed alkaline. A more convenient scale for acidity and alkalinity results from taking the logarithm of the reciprocal of the hydrogen ion activity (or concentration), this is the pH of the solution.
Neutralization and hydrolysis	Neutralization is the result of the reaction between equivalent amounts of an acid and base. The extent to which the reaction proceeds depends not only on the strength of the acid and base, but also on the nature of the solvent. It is possible that the solvent may interact with the products of neutralization, thus reversing it. This process has the general term solvolysis, and is known as hydrolysis when the solvent is water.
Diprotic and polyprotic acids	Many acids have the capacity to lose more than one proton. As would be expected following each deprotonation the conjugate base becomes stronger and the reverse reaction more favored, that is the dissociation constant (K_a) for successive deprotonations decreases and distinct values for K_a may be determined.
Related topics	Water – the biological solvent (C) Dissociation constants of acids and Lowry–Bronstead acid and base bases (N2) (N1)

---

pH

The (solvated) hydrogen ion (H^+) and hydroxide ion (^-OH) produced by the self-ionization of pure water must be present in equivalent amounts, thus water is neither acidic nor basic, but rather is referred to as being **neutral**. As the **ionic product** of water (K_w) (see Topic N2) is 10^{-14} M^2 at room temperature, the concentration for each of the ions must be the square root of this number (i.e. 10^{-7} M).

If a source of H^+ or ^-OH is present other than the water itself, that is the ions are present at a concentration greater than 10^{-7} M, then the solution is **acidic**. Conversely, if the concentration of H^+ is less than 10^{-7} M (hence [^-OH] is greater than 10^{-7} M) the solution is referred to as **alkaline** or **basic**. There is, however, a more convenient way of expressing the acidity or basicity (alkalinity) of a solution and that is by using the **pH** scale. The pH of a solution is simply the logarithm (to the base 10) of the reciprocal of the hydrogen ion activity, which under conditions of dilute solutions may be approximated to the concentration, as follows;

$$pH = -\log [H^+]$$

Therefore,

$$pH + pOH = pK_w$$

Of course, as both K_a and K_b (see Topic N2) are related to the ionic product the pH scale may also be applied to these terms. Thus;

$$pK_a + pK_b = pH + pOH = pK_w$$

The pH scale is thus a convenient system for specifying the acidity of a solution in terms of small, and for the most part, positive numbers, over the range 0 to 14, in general.

It must be noted that as the ionic product of water varies with temperature, so too does the pH. Hence, when quoting pH values the temperature for the measurement must be specified.

Neutralization and hydrolysis

The original theory of neutralization, described by Arrhenius, stated that neutralization was the result of the reaction between an acid and base to produce a salt and water. Extending this to the more general situation when water may not necessarily be involved, neutralization may be described as follows;

$$\underset{\text{acid}}{HA} + \underset{\text{base}}{B} \rightleftharpoons BH^+ + A^-$$

The extent to which the neutralization reaction proceeds depends not only on the strength of the acid and base components but also on the nature of the solvent. If the solvent is **amphiprotic** (i.e. able to act as a base or an acid) as is water for example, it may react with the products of neutralization as follows;

$$BH^+ + H_2O \rightleftharpoons B + H_3O^+ \quad \text{(I)}$$
$$A^- + H_2O \rightleftharpoons HA + {}^-OH \quad \text{(II)}$$

The neutralization reaction is consequently reversed. This general process is referred to as **solvolysis**, and more specifically when water is the solvent, **hydrolysis**.

Reaction (I) is favored when the base is weak, and reaction (II) when the acid is weak. Under these conditions hydrolysis is favored. However, if both the acid and base are strong then hydrolysis is not favored, and neutralization will occur when they are present in equal amounts. In contrast, a solution of equivalent amounts of strong acid and weak base (favoring reaction (I)) or strong base and weak acid (favoring reaction (II)) will result in the formation of acidic and alkaline solutions, respectively. Clearly when equivalent amounts of a weak acid and weak base react, as both hydrolysis reactions are favored, the acidity or alkalinity of the resulting solution will depend on the relative strengths of the acid and base.

Diprotic and polyprotic acids

A number of (biochemically) important acids have the ability to lose several protons, in a step-wise fashion. An example of such is phosphoric acid, H_3PO_4. In aqueous solution phosphoric acid may undergo the following dissociation reactions, each have an associated dissociation constant;

$$H_3PO_4 + H_2O \rightleftharpoons H_3O^+ + H_2PO_4^- \qquad K_a = 7.1 \times 10^{-3}$$
$$H_2PO_4^2 + H_2O \rightleftharpoons H_3O^+ + H_2PO_4^{2-} \qquad K_a = 6.3 \times 10^{-8}$$
$$HPO_4^{2-} + H_2O \rightleftharpoons H_3O^+ + PO_4^{3-} \qquad K_a = 4.7 \times 10^{-13}$$

It is clear that the equilibria commences with H_3PO_4 as an acid with its conjugate base $H_2PO_4^-$. However, in the second equilibrium process $H_2PO_4^-$ is now the acid. The decrease in the dissociation constants clearly indicates the difficulty of successive deprotonations.

N4 BUFFERS

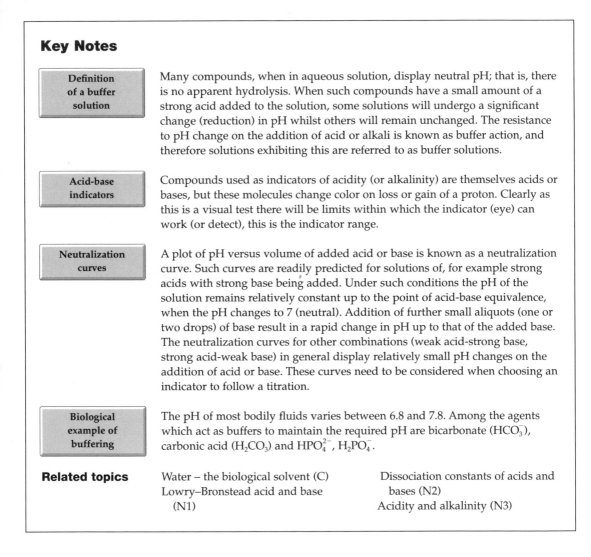

Key Notes

Definition of a buffer solution

Many compounds, when in aqueous solution, display neutral pH; that is, there is no apparent hydrolysis. When such compounds have a small amount of a strong acid added to the solution, some solutions will undergo a significant change (reduction) in pH whilst others will remain unchanged. The resistance to pH change on the addition of acid or alkali is known as buffer action, and therefore solutions exhibiting this are referred to as buffer solutions.

Acid-base indicators

Compounds used as indicators of acidity (or alkalinity) are themselves acids or bases, but these molecules change color on loss or gain of a proton. Clearly as this is a visual test there will be limits within which the indicator (eye) can work (or detect), this is the indicator range.

Neutralization curves

A plot of pH versus volume of added acid or base is known as a neutralization curve. Such curves are readily predicted for solutions of, for example strong acids with strong base being added. Under such conditions the pH of the solution remains relatively constant up to the point of acid-base equivalence, when the pH changes to 7 (neutral). Addition of further small aliquots (one or two drops) of base result in a rapid change in pH up to that of the added base. The neutralization curves for other combinations (weak acid-strong base, strong acid-weak base) in general display relatively small pH changes on the addition of acid or base. These curves need to be considered when choosing an indicator to follow a titration.

Biological example of buffering

The pH of most bodily fluids varies between 6.8 and 7.8. Among the agents which act as buffers to maintain the required pH are bicarbonate (HCO_3^-), carbonic acid (H_2CO_3) and HPO_4^{2-}, $H_2PO_4^-$.

Related topics

Water – the biological solvent (C)
Lowry–Bronstead acid and base (N1)

Dissociation constants of acids and bases (N2)
Acidity and alkalinity (N3)

Definition of a buffer solution

Aqueous solutions of sodium chloride and of ammonium acetate are neutral; they display pH 7. Sodium chloride is the salt of a strong acid (HCl) and a strong base (NaOH), hence hydrolysis does not occur (see Topic N3). In contrast ammonium acetate is the salt of a weak acid and a weak base; as their strengths are almost equivalent such a solution is almost neutral (see Topic N3). However, the differing nature of the acids and bases in the two solutions (that is their differing responses to neutralization) means that they respond differently to the addition of small quantities of acid or base. If a small amount of hydrochloric acid is added to the sodium chloride solution the pH of the solution will drop markedly. If the same concentration of acid is added to a solution of ammonium acetate however, the pH remains virtually unchanged. Ammonium acetate is thus said to display

buffer action, and the solution of ammonium acetate is known as a **buffer solution**.

In general, if a buffer solution is made up of a mixture of a weak acid and one of its salts then the buffering action may be described as follows;

$$H_3O^+ + A^- \longrightarrow HA + H_2O \tag{1}$$

or

$$^-OH + HA \longrightarrow A^- + H_2O \tag{2}$$

In (1) when hydrogen ions are added, they are removed by combination with the anion of the salt to form the un-ionized acid (HA). In (2) the effect of adding hydroxide ions is illustrated.

If the buffer, on the other hand, is made up of a weak base and one of its salts, the buffering action may be described as follows;

$$H_3O^+ + B \longrightarrow BH^+ + H_2O \tag{3}$$

or

$$^-OH + BH^+ \longrightarrow B + H_2O \tag{4}$$

Taking account of the fact that if an acid or base is weak it will not appreciably dissociate in solution, the dissociation constant (see Topic N2) for each of the buffer solutions referred to may be expressed as follows;

$$K_a = \frac{[H^+][salt]}{[acid]} \text{ or } K_b = \frac{[^-OH][salt]}{[base]}$$

Where [acid] and [base] refer to the total concentrations of acid and base, respectively, and [salt] refers to the concentration of A^- or BH^+.

Rearranging these expressions and taking logarithms, the following expressions arise;

$$pH = pK_a + \left(\frac{log\,[salt]}{[acid]}\right) \qquad pOH = pK_b + \left(\frac{log\,[salt]}{[base]}\right)$$

Where pH, pK_a, and pK_b refer to the negative logarithm (to the base 10) of $[H^+]$, K_a, and K_b, respectively.

These are often referred to as the **Henderson–Hasselbalch** equations and are reasonably accurate within the pH range 4–10.

In addition to knowing that a solution may act as a pH buffer, it is also important to know the **buffering capacity**. That is the amount of acid or base the solution can absorb before a pH change is brought about. A rule of thumb guide is that it is generally the case that buffer solutions have their greatest capacity when the ratio of salt to acid, or salt to base is one.

Some examples of buffers and their ranges are provided in *Table 1*.

Table 1. Some buffer solutions and their ranges at 25°C

Solution	pH Range
Boric acid; borax	6.7–9.2
Acetic acid; sodium acetate	3.7–5.6
KH_2PO_3; K_2HPO_3	5.3–8.0

Acid-base
indicators

If the concentration of, for example, an acid (or acid component of a molecule) is unknown it may be determined by adding aliquots of a known concentration of a base. At the equivalence point (between acid and base) the pH of the solution is neutral. Such a reaction could be monitored by use of a pH meter, alternatively a pH or acid-base indicator could be used. Compounds used as indicators of acid-base reactions are routinely weak organic acids or weak organic bases; the un-ionized forms of which generally exist as tautomeric mixtures (see Topic J2), which have differing propensity to ionize and differ in color;

$$HIn_a + HIn_b \rightleftharpoons H^+ + In_b$$
$$\text{(color a)} \qquad\qquad\qquad \text{(color b)}$$

HIn_a is one tautomer of indicator acid (a) with tautomer b (HIn_b) being more readily ionized.

Clearly if acid is added to a solution of this indicator the equilibrium will lie more to the left, and the solution will display the color a. In contrast, if alkali is added the equilibrium will shift and the solution will exhibit color b.

An example of an indicator which behaves in this way is provided by phenolphthalein (see *Fig. 1*).

As two steps are involved, expressions may be written for two equilibirum contants; K, and K' representing the tautomeric and ionization equilibria, respectively. The product of these two constants is K_i, the **indicator constant**, and the **indicator equation** is as follows;

$$K_i = \frac{x}{1 - x}[H^+]$$

where x is the fraction of the indicator in the form leading to color b. Similar expressions may be written when the indicator is a weak base.

The expression above may be rearranged and utilized to determine the **working range** of the buffer; that is the pH range over which a color change may take place.

$$pH = 21 + pK_i \quad \text{to show acid color}$$
$$pH = 1 + pK_i \quad \text{to show base color}$$

Tautomers

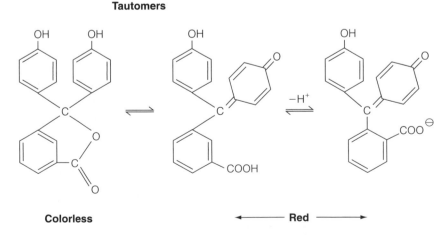

Colorless   ◄——— Red ———►

Fig. 1.   Tautomeric forms of phenolphthalein.

Thus, the complete range over which a color change may be detected is;

$$pH = pK_i \pm 1.$$

In *Table 2* some examples of indicators and their working ranges are provided.

Table 2. Some examples of indicators and their working ranges

Indicator	Color		pH Range
	Acid	Base	
Thymol blue	Red	Yellow	1.2–2.8
Bromophenol blue	Yellow	Blue	3.0–4.6
Cresol red	Yellow	Red	7.2–8.8
Methyl orange	Orange	Yellow	3.1–4.4
Phenolphthalein	Colorless	Pink	8.3–10.0

Neutralization curves

As indicated in *Table 2*, the pH range over which the color of an indicator might change is approximately 2 units. Clearly if the pH of a solution of indicator changes only slowly over this range, then so will the color change slowly. In contrast, if the pH changes rapidly, a sharp color change, which is generally more readily detected, will result. Clearly the rate of change of pH during a titration, particularly close to the **equivalence** point (when acid and base concentrations are equivalent), is of prime importance in determining the accuracy of the titration. In order therefore to determine which indicator is best for the acid-base process of interest, it is important to consider the pH range over which neutralization occurs. To do this neutralization curves, plots of pH versus volume of (for example) added alkali, should be predicted. In *Fig. 2a* and *2b* the neutralization curves for 25 cm^3 volumes of 0.1 M concentrations of acid and alkali solutions, respectively, are shown. In the case of a mixture of strong base and strong acid, the addition of 0.1 cm^3 of the same concentration of base (that is approximately two drops) results in a pH change of 6 units. This would result in a sharp color change for any indicator working in the region 4–10 pH units. In contrast, in the case of a weak acid and a strong base, the neutralization curve is much shallower, covering a range of just 2 pH units. In order to get a sharp end point for this titration, an indicator working in the range 7.8–10 pH units must be selected.

Biological example of buffering

The pH of human body fluids varies greatly depending on location. For example the pH of blood plasma is around 7.4, whereas that of the stomach is around 1. Although there is this variation, the pH of each type of body fluid must be maintained by a series of buffering processes. As an illustration as to how buffering comes about, the control of blood pH is considered.

The body has two key methods for handling the acid produced by metabolic processes and thus prevent pH lowering, one of these is buffering. Blood consists essentially of two components, plasma and red blood cells (or erythrocytes). Blood plasma is maintained at pH 7.4, largely by the buffer systems HCO_3^-/H_2CO_3 and $HPO_4^{2-}/H_2PO_4^-$, and various plasma proteins (the amino acids of which can also display buffering action (see Topic M1)), in particular albumin. Hemoglobin (HHb), located in red blood cells, together with its oxygenated counterpart, oxy-hemoglobin (HHbO$_2$), also play a significant role in buffering. Both of these proteins are weak acids;

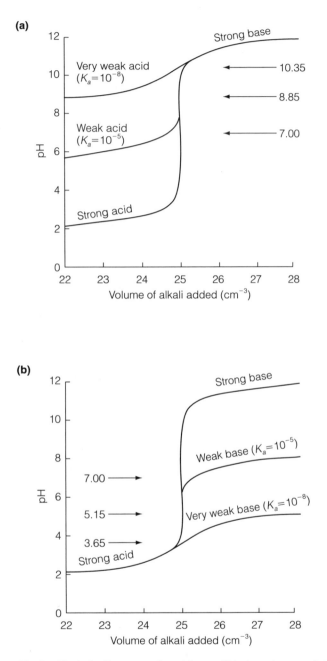

Fig. 2. Neutralization curves for mixtures of (a) strong base and strong acid or weak acid, and (b) strong acid and strong base or weak base, as a function of added alkali. The numbers at the side of the curves refer to the pH at the equivalence point.

$$HHb \rightleftharpoons H^+ + Hb^- \qquad pK_a = 8.2$$
$$HHbO_2 \rightleftharpoons H^+ + HbO_2^- \qquad pK_a = 6.95$$

A cascade of equilibrium reactions occur involving all of these components and carbon dioxide (CO_2). A summary of the events which occur is provided in *Fig. 3*.

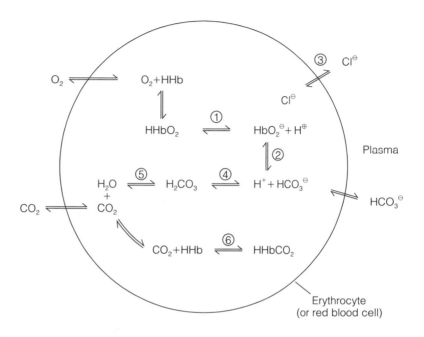

Fig. 3.   The role of hemoglobin, oxyhemoglobin, carbonic acid, and bicarbonate in maintaining the pH of blood plasma. ① Deprotonation of oxyhemoglobin; ② proton is used with bicarbonate to form carbonic acid; ③ chloride transport; ④ formation of carbonic acid; ⑤ dissociation of carbonic acid to form water and carbon dioxide; ⑥ formation of carbaminohemoglobin.

N5 SOLUBILITY

Key Notes

Solubility product	Most ionic compounds are soluble in water, but a number of inorganic salts are actually only weakly soluble in water. For a given salt M_xA_y, solubility is limited by the value of its particular solubility product K_{sp}, given by $[M]^x[A]^y$. The compound will dissolve until the ion concentrations are such that K_{sp} is reached. At this stage further addition of M_xA_y produces a precipitate of the salt.
Common ion effect	The solubility of a salt can be significantly reduced in the presence of another salt, with which it has one ion species in common. This is the common ion effect.
Insoluble hydroxides	The solubility of a metal cation can become significantly reduced by the presence of particular anions in solution. Importantly, metal hydroxides may readily precipitate, especially if the pH is above 7. Thus, a number of metal ions are only ever found within complexes in biological systems, rather than as free ions.
Fat-soluble molecules	There are many physiologically important molecules which are not water-soluble, such as the fat-soluble vitamins. These generally reside in hydrophobic niches, and can be transported by lipoproteins in the blood plasma.
Related topics	Iron (G2) Acidity and alkalinity of aqueous Hydrophobic interactions (H2) solutions (N3) Water – the biological solvent (C) Buffers (N4)

Solubility product   Many biochemical processes occur in aqueous solution, which means that the various participants must be adequately water-soluble. If this is not the case, a new phase can form, such as a solid precipitate, or a layer of hydrophobic oil. This is biologically undesirable, because reactions between molecules in different phases are generally slow and inefficient. Nature has evolved several strategies to circumvent such problems.

Although ionic solids generally dissolve in water, this is not universally true. A number of inorganic salts have only limited solubility in water. The reasons for this can arise from the favorable stabilizing forces that exist within the solid compound, set against those among the dissolved ions and water molecules. This can mean that the amount of such a salt which can be dissolved, depends on factors such as the pH (see Topic N3) and the presence of other dissolved salts.

For such an ionic compound, the maximum amount of it which can be in solution is dictated by its **solubility product**, a principle established by Nernst in 1889. Thus, for a salt composed of metal cation M, and anion A, with the formula M_xA_y, the dissolution of which can be represented by;

$$M_xA_{y \; (solid)} \longrightarrow xM_{(aqueous)} + yA_{(aqueous)}$$

The solubility product, K_{sp}, of this salt is given by the product $[M]^x[A]^y$, and is (essentially) constant for a given salt. For highly soluble salts, such as NaCl, this number is relatively large, and solubility only becomes a consideration at salt levels of several moles per liter. With decreasing solubility, K_{sp} becomes smaller. This product represents the maximum solubility of the salt. As an example, for AgCl, K_{sp} is 10^{-10} at 25 °C. In a saturated solution and with no other salts present, $[Ag^+] = [Cl^-]$ and so the (maximum) concentration of silver ions in solution will be the square root of K_{sp} (i.e. 10^{-5} M). Addition of further AgCl will not increase this, but will form a **precipitate** instead. Now consider when AgCl is added to a solution which already contains 100 mM NaCl, that is, in which $[Cl^-]$ is 10^{-1} M. The amount of extra Cl^- ions from added AgCl can be regarded as neglible compared to this, so now;

$$[Ag^+] \times 10^{-1} = K_{sp} = 10^{-10}$$

Thus the concentration of the silver ions in a solution is a mere 1×10^{-9} M, i.e. 1 nM, which is significantly reduced from before.

Common ion effect

Following on from the definition of the solubility product, the solubility of silver chloride can be said to have been depressed by a **common ion effect**. At low ionic strengths, the actual concentration of ions can be used in solubility calculations without serious error. At higher ionic strengths the activities (see Topic N1) should be used instead, although calculations based on concentrations may still be useful as qualitative guides.

In some cases, a particular salt may itself be fairly soluble, but new combinations of ions, corresponding to less soluble salts, may form when it is added to a solution. For example, $AgNO_3$ is very soluble, but when added to a solution of NaCl, a precipitate of AgCl is likely.

Insoluble hydroxides

Common ion effects must also be taken into consideration when talking about the formation of weakly soluble metal hydroxides (see Topics G2 and G3).

Consider the addition of $ZnCl_2$ to a solution which is well buffered (see Topic N4) at pH 7. Although $ZnCl_2$ itself is fairly soluble, it will not be the least-soluble zinc salt present, since $Zn(OH)_2$ has limited solubility;

$$[Zn^+] [OH^-]^2 = 4 \times 10^{-16}$$

At pH 7, $[OH^-]$ is 10^{-7} M (see Topic N3), so the maximum concentration of Zn^{2+} ions in solution will be $K_{sp}/\{(10^{-7})^2\}$, which is 0.04 M. However, if the buffer is set at pH 8, so $[OH^-]$ is now 10^{-6} M, the Zn^{2+} concentration is now limited to only 0.0004 M. Thus, the pH has a strong effect on the solubility of Zn^{2+}. For iron (both as Fe^{2+} and Fe^{3+}, see Topic G2), solubility of the hydroxides is even more limiting than for Zn^{2+}. However, physiologically the iron ions can be protected from precipitation by being strongly complexed within specialized proteins (e.g. transferrins, see Topic G2) and thus transported without risk of precipitation. This strategy is used widely in nature, and is why such ions are not generally encountered at any significant level as being 'free' in solution within living tissues.

Fat-soluble molecules

Although overall living tissue may be regarded as being essentially an aqueous environment, this is a simplification. For example, fat globules or deposits may lie within or around cells. We have the lymphatic system, which conveniently transports material that would otherwise clog the blood system. The cells which comprise living tissue are surrounded by a plasma membrane (see Topic H2), which can be regarded as hydrophobic regions. Significant components of these membranes are lipoproteins, which are associations of proteins with lipids (see Topics H2 and K3). Units of these also exist suspended in the blood plasma, where they are important in the transport of molecules which are not soluble in water.

Examples of important compounds which are not soluble in water can be found amongst the vitamins. Vitamins can be classed into two broad groups: those which are water soluble, and those which are not. The latter include vitamins A, D, E and K, which are all nonpolar hydrocarbons and will dissolve instead in hydrophobic media, and so are often referred to as the **fat-soluble vitamins** see Topic K3). These are transported in the bloodstream as 'passengers' of lipo-proteins.

01 BASIC CONCEPTS

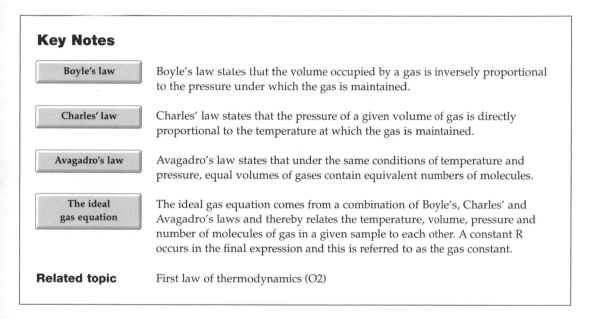

Key Notes

Boyle's law	Boyle's law states that the volume occupied by a gas is inversely proportional to the pressure under which the gas is maintained.
Charles' law	Charles' law states that the pressure of a given volume of gas is directly proportional to the temperature at which the gas is maintained.
Avagadro's law	Avagadro's law states that under the same conditions of temperature and pressure, equal volumes of gases contain equivalent numbers of molecules.
The ideal gas equation	The ideal gas equation comes from a combination of Boyle's, Charles' and Avagadro's laws and thereby relates the temperature, volume, pressure and number of molecules of gas in a given sample to each other. A constant R occurs in the final expression and this is referred to as the gas constant.
Related topic	First law of thermodynamics (O2)

Boyle's law

In 1662 Boyle's study of the behavior of gases led him to determine that the volume (V) occupied by a gas, **at constant temperature**, was inversely proportional to the pressure (P);

$$V \propto 1/P$$
$$\text{or} \quad PV = constant$$

This expression is known as Boyle's law. A plot of **P** versus **V** is a hyperbola, as this is at a specific temperature it is referred to as an **isotherm**.

Boyle's law is thus used to predict the pressure of a gas when its volume is changed and vice versa.

Boyle's studies further showed that at constant pressure the volume of a given amount of gas is directly proportional to its temperature (T);

$$V \propto T$$
$$\text{or} \quad V/T = constant$$

Charles' law

More than a century after Boyle's pioneering work, Charles in 1787 (and later in greater detail Gay-Lussac in 1802) in his quantitative study of the thermal expansion of gases, determined that at a constant volume, the pressure of a given amount of gas is directly proportional to its temperature;

$$P \propto T$$
$$\text{or} \quad P/T = constant$$

This is Charles' law.

Avagadro's law

Shortly after the work of Gay-Lussac, Avagadro formulated a third law, which proposed that under the same conditions of temperature and pressure, equal volumes of gases contain the same number of molecules;

$$V/n = constant$$

where **n** is the **number of molecules**. This expression is therefore known as Avagadro's law.

The ideal gas equation

The ideal gas equation occurs from a combination of Boyle's, Charles', and Avagadro's laws and is the fundamental expression for describing the properties of gases under ideal conditions.

If the volume, V, of a gas is considered then the three laws state the following;

$$
\begin{aligned}
V &\propto 1/P \quad \text{(at constant T and n)} \quad \text{(Boyle)} \\
V &\propto T \quad\ \ \text{(at constant P and n)} \quad \text{(Charles)} \\
V &\propto n \quad\ \ \text{(at constant T and P)} \quad \text{(Avagadro)}
\end{aligned}
$$

Then V must be proportional to the product of these three terms, that is;

$$V \propto nT/P$$

The proportionality constant is now written as **R**, the **gas constant**, and the expression rearranged such that;

$$PV = nRT$$

This is the ideal gas equation.

02 FIRST LAW OF THERMODYNAMICS

Key Notes

Heat and work	The classical description of work is, that it is the product of force and distance. In thermodynamics the concept of 'work' is more widely based; including gravitational effects, electrical changes, and the act of expansion of a gas. Thus, work should really be thought of as a mode of energy transfer. Another mode of energy transfer is called heat. Neither of these processes are so-called 'state functions' as each of them depends on how the change of state (be it electrical, gravitational or thermal) occurs, rather than just the initial and final conditions.
Enthalpy	The internal energy of a system is a state function since it depends only on the thermodynamic properties of the state; that is, temperature, pressure, and composition. The change in internal energy of a system is related to the heat change of the system and the change in work. In the laboratory, reactions are generally carried out at constant pressure and therefore it is possible to write an expression for the internal energy of the system in terms of pressure, volume and heat change. In doing so, a new function may be introduced called enthalpy (H); this is equal to the internal energy of the system plus the product of pressure and volume. Enthalpy is therefore also a function of state.
Heat capacity	When heat is added to a substance its temperature will rise. The extent to which the temperature rises depends on the amount of heat applied, the amount of substance present, the chemical and physical nature of the substance and the conditions under which heat is added. In short, the temperature rise for a given amount of substance is proportional to the heat added. The proportionality constant is called the heat capacity.
Enthalpy of formation	The principle of enthalpy can be applied to chemical reactions which are almost always accompanied by heat change. The enthalpy of formation is defined as the sum of the change of the enthalpy changes for the products of a reaction minus the sum of enthalpy changes for the reactants.
Dependence of enthalpy on temperature	The change of enthalpy in a chemical reaction is dependent on the heat capacity of the components of the reaction and the temperature change involved for the products and reactants involved. An expression relating enthalpy to temperature was devised by Kirchoff.
Related topic	Basic concepts (O1)

Heat and work   The concept of **work** is very subtle but widely based. As shown in *Table 1*, the thermodynamic notion of work covers not only mechanical work, but also surface, electrical, gravitational, and expansion work. Work is therefore one more type of energy transfer. In the present context, an important example of a system doing work, is that of an expanding gas.

Table 1.   Definitions of the various types of work covered in thermodynamics

Type of work	Expression	Symbol meaning
Mechanical	f dx	f, force; dx, distance traveled
Surface	γ dA	γ, surface tension; dA, change in area
Electrical	ε dQ	ε, potential difference; dQ, change in electric charge
Gravitational	Mg dh	M, mass; g, acceleration due to gravity; dh, change in height
Expansion	P dV	P, pressure; dV, change in volume

Consider a system in which gas is contained in a cylinder and is kept in there by a frictionless piston. Assuming there is no external pressure, the work (W) involved in expanding the gas will simply be that of moving the mass of the piston (M) over a distance (Δh);

$$W = -\text{force} \times \text{distance}$$
$$= -\text{mass} \times \text{acceleration} \times \text{distance}$$
$$= -Mg\ \Delta h$$

where g is the acceleration due to gravity. Note that the work done is a **negative** quantity. This is always the case when work is done against the surroundings. In contrast, compression involves a **positive** amount of work (see *Table 2*).

Table 2.   Sign conventions for work and heat

Process	Sign
Work done by system on surroundings	−
Work done on system by surroundings	+
Heat absorbed by system from surroundings (endothermic)	+
Heat absorbed by surroundings from system	−

If, as in a more realistic case, an external pressure is applied to the piston, then it is possible to modify the expression for work as follows (see Topic O1);

$$W = -P_{ex}\ \Delta V$$

where P_{ex} is the external pressure and ΔV the volume change during expansion. Following on from this and utilizing the ideal gas equation (see Topic O1), it is possible to write another expression for the work involved in making volume changes during expansion;

$$W = -nRT \ln (V_2/V_1)$$

V_1 and V_2 are the initial and final volumes, **n** is the number of molecules of gas, **R** is the gas constant and **T** the temperature. This expression applies to **reversible** processes, however real processes are always **irreversible**. Nevertheless dealing with reversible processes enables us to calculate the **maximum** amount of work that could be extracted. This quantity is important in estimating the efficiency of chemical and biochemical processes.

Another mode of energy transfer involves the quantity **heat**. Heat flows from a hot object to a cold object. If the temperatures of the two objects were equal the word heat is not applicable. Like work, heat is not a function of state, as the processes involved in exchanging heat are important, not just the initial and final states.

When discussing heat it is more usual to talk in terms of a **heat change** and this is symbolized by **Q**.

Enthalpy

Unlike heat and work, the **internal energy (U)** of a system is a function of state; it depends on other functions of state such as temperature, pressure, and composition.

When talking about internal energy, we are accounting for **internal vibrations, rotations, translations, electronic energy** and **nuclear energy**. We are neglecting external factors such as kinetic energy and potential energy, by assuming that the system is at rest and that there is no external electrical or magnetic field present. Under these conditions the total energy of the system is equal to the internal energy. We have no way of calculating the internal energy accurately, so we restrict ourselves to talking about energy differences (ΔU). This leads to the **first law of thermodynamics**. The internal energy change of a system is equal to the heat change and the work performed on changing its condition;

$$\Delta U = Q + W$$

Encompassed in this expression is the principle of **conservation of energy**; energy can neither be created nor destroyed, it is simply converted from one form into another. As under laboratory conditions we generally work at constant pressure, it is possible to modify this expression as follows;

$$\Delta U = Q_p - P\Delta V$$

where Q_p is the heat change at pressure P. Rearranging this expression and considering a volume change from V_1 to V_2 then Q_p may be expressed as;

$$Q_p = (U_2 + PV_2) - (U_1 + PV_1)$$

This leads to the definition of another function called **enthalpy (H)** as follows;

$$H = U + PV$$

At constant pressure therefore;

$$\Delta H = \Delta U + P\Delta V$$

Clearly as enthalpy depends on functions of state alone it is also a function of state.

Heat capacity

Most chemical reactions are accompanied by either evolution (**exothermic reactions**) or absorption (**endothermic reactions**) of heat. The enthalpy change for any process involving heat depends on the temperature change involved and the **heat capacity** (C) of the reactants and products (see next page).

In general the temperature rise in a given amount of substance is directly proportional to the heat added;

$$\Delta T \propto Q$$

alternatively

$$C = \frac{Q}{\Delta T}$$

At constant volume the heat capacity may be expressed as;

$$C_v = \frac{Q_v}{\Delta T} = \frac{\Delta U}{\Delta T}$$

and at constant pressure as follows;

$$C_p = \frac{Q_p}{\Delta T} = \frac{\Delta H}{\Delta T}$$

Enthalpy of formation

In general, chemical reactions are either **exothermic** (heat is given off to the surroundings), or **endothermic** (heat is absorbed from the surroundings). In the first instance ΔH is negative, in the latter ΔH is positive.

Consider the reaction between carbon (in the form of graphite) and oxygen;

$$C\,(s) + O_2(g) \rightarrow CO_2(g)$$

When one mole of carbon is reacted with one mole of oxygen at 1 atm pressure and 298 K, one mole of carbon dioxide is produced with 393.51 kJ of heat given off. That is, $\Delta H = -393.51$ kJ. If we could determine the enthalpies of carbon, oxygen and carbon dioxide we could calculate ΔH, but as this is not possible we always need to deal with enthalpy differences. The enthalpies of each of the elements are arbitrarily set at zero.

A value for ΔH has been quoted for this reaction, however as for the most part we are talking about enthalpy changes at 298K (25 °C) and 1 atm, the symbol ΔH° is often used. In this particular example the enthalpy of carbon dioxide is equivalent to ΔH° which we will now call ΔH_f°, the enthalpy of formation.

In general the standard enthalpy of a reaction may be expressed as follows;

$$\Delta H^{\circ} = \Sigma\ \Delta H_f^{\circ}(\text{products}) - \Sigma\ \Delta H_f^{\circ}(\text{reactants})$$

It should be noted, that if more than one step is involved in a reaction process then the overall enthalpy change for the reaction, is the sum of the enthalpy changes for each of the individual steps.

Dependence of enthalpy on temperature

For any reaction the enthalpy change is given by;

$$\Delta H = H_{\text{products}} - H_{\text{reactants}}$$

If this equation is differentiated with respect to temperature, at constant pressure then we get to an expression in terms of heat capacities;

$$\frac{d\ \Delta H}{dT} = \frac{dH_{\text{products}}}{dT} - \frac{dH_{\text{reactants}}}{dT}$$

$$= C_p(\text{products}) - C_p(\text{reactants}) = \Delta C_p$$

Upon integration this yields the following;

$$\Delta H_2 - \Delta H_1 = \Delta C_p\ (T_2 - T_1)$$

This is known as Kirchoff's equation.

O3   SECOND LAW OF THERMODYNAMICS

Key Notes

Spontaneous processes

The first law of thermodynamics is concerned with the conservation of energy during a process of change, it says nothing about whether a process will actually occur. This is dealt with by the second and third laws with the introduction of the terms entropy (S) and Gibbs free energy (G). To understand why further thermodynamic parameters are required, it is necessary to understand processes which occur of their own accord, that is, spontaneously. For example ice melts at 20 °C but water at the same temperature will not spontaneously form ice. Consideration of energy change cannot predict the direction of a spontaneous process therefore, a further thermodynamic parameter is required, and this is entropy.

Statistical definition of entropy

The statistical definition of entropy comes from consideration of the probability of gaseous molecules occupying particular volumes of space.

Thermodynamic definition of entropy

It is not convenient to determine probabilities for complex, macroscopic (real) systems. However, the expression determined from the statistical treatment of spontaneous processes can readily be related to other thermodynamic parameters.

Temperature dependence of entropy

When the temperature of a system is raised, the entropy of the system also increases. The entropy change with temperature may be determined for a constant pressure system provided the heat capacity of the system is also known.

Related topics   Basic concepts (O1)   First law of thermodynamics (O2)

Spontaneous processes

In everyday life, a huge number of spontaneous processes occur. Ice melts at 20 °C, a leaf falls from a tree to the ground, but the reverse processes do not happen spontaneously. Indeed, in general it can be said that if a process occurs spontaneously in one direction it cannot also occur spontaneously in the opposite direction. In considering what changes accompany spontaneous processes it is logical to assume that a **spontaneous process** occurs so as to **lower the energy** of the system. This certainly helps explain why things fall and do not rise again spontaneously. However, this is not the complete answer. It cannot explain why, for example, the expansion of an ideal gas, at constant temperature, does not involve a change in its internal energy (see Topics O1 and O2). Also, when ice melts the internal energy of the system actually increases. Hence, as energy change cannot predict the direction of spontaneous reactions, a further thermodynamic parameter is required, this is **entropy (S)**.

Statistical definition of entropy

In trying to understand spontaneous processes it is helpful to consider systems involving a large number of molecules and to apply a statistical treatment to the behavior of these. But we can start off with small numbers of molecules. Thus, consider a cylinder into which gas molecules are being introduced (see *Fig. 1*). With one molecule present, and the volume within the cylinder being V_2, then the probability (ω) of finding the molecule in V_2 is obviously 1 (100%). Likewise, the probability of finding it in one particular half of the cylinder (a volume of V_1) is 0.5 (50%). If a second molecule is introduced, the probability of finding both molecules in V_2 is still 1, but the probability of finding both in V_2 is 0.25. This is because of the probability of 0.5 arising from the first molecule, multiplied by 0.5 for the second. If we add a third molecule, the probability of finding all three in V_2 is 0.125 (i.e. 0.5^3). Thus, when we get to more realistic numbers of molecules, ω for finding all the molecules in V_1 becomes extremely small. So, if the volume of the cylinder is reduced to V_1 and then increased to V_2, it is highly unlikely that the gas molecules will all remain in V_1. This analysis shows how probability is an important factor in describing a system, and predicting how it will behave. It is therefore reasonable to associate **entropy** with probability. The appropriate expression involves the natural logarithm of the probability:

$$S = k \ln \omega$$

This is known as the **Boltzmann equation**, and k is the Boltzmann's constant. (Boltzmann was a 19th century physicist, whose ideas were not generally accepted until after his death in 1906.) If we have a process which proceeds from state 1 (which has a probability of ω_1) to state 2 which has a probability of ω_2, then the change in entropy is:

$$\Delta S = k \ln (\omega_2) - k \ln (\omega_1) = k \ln (\omega_2/\omega_1)$$

This expression may be modified, as probabilities may be expressed in terms of volumes, eventually leading to;

$$\Delta S = n R \ln (V_2/V_1)$$

where n is the number of moles of the gas present, R is the **ideal gas constant**.

It should be noted that this expression holds at constant temperature and also

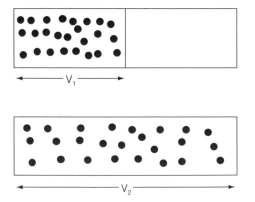

Fig. 1.   Schematic representation of a container of N molecules occupying volumes V_1 and V_2.

that it is not necessary to specify the manner in which change has been brought about. This is because **entropy** is another **function of state**.

The Boltzmann equation leads us to the **third law of thermodynamics**. This states that the entropy of a perfect crystal is zero at a temperature of absolute zero. That is to say: at this temperature the crystal is perfectly ordered, so only one arrangement of the constituent atoms is possible (neglecting any interchanging of identical atoms), and so its probability is 1, and $\ln(\omega)$ is zero. This is a somewhat abstract concept, since a temperature of absolute zero has never been achieved experimentally. In fact, it is unlikely that absolute zero ever can be achieved, though it may be closely approached.

For completeness, we can add another thermodynamic 'law'. If two objects are in thermal equilibrium with a third object, they will be in thermal equilibrium with each other. This is sometimes called the **zeroth law of thermodynamics**, and is a way of defining temperature.

Thermodynamic definition of entropy

It is not convenient to calculate probabilities for real systems, fortunately this is not necessary as entropy may be determined from the consideration of other thermodynamic parameters.

It has previously been shown (see Topic M2) how the heat absorbed by an ideal gas may be related to a change in volume. Hence, it is possible to write an expression for entropy which is dependent on heat;

$$\Delta S = \frac{Q_{rev}}{T}$$

where Q_{rev} is the heat absorbed during a reversible process.

So far the entropy change of the system alone has been considered but if this treatment is extended to the surroundings it becomes apparent that **the entropy of an isolated system increases in an irreversible process and remains unchanged in a reversible process. It can never decrease**. This is the second law of thermodynamics.

Temperature dependence of entropy

When the temperature of a system is raised from T_1 to T_2 its entropy also increases. If an amount of heat is transferred to the system in a reversible manner then the entropy change for the system may be written as follows;

$$S_2 = S_1 + \Delta S$$

However, $dQ_{rev} = dH$, and $dH = C_p dT$, thus;

$$S_2 - S_1 = \Delta S = C_p \ln T_2/T_1$$

04 GIBBS FREE ENERGY

Key Notes

Gibbs free energy

The first law of thermodynamics is concerned with energy balance and the second law helps us decide whether a process can occur spontaneously. However, in applying the second law and using the concept of entropy, we have to consider both the system and the surroundings. So that only the system need be considered, a third thermodynamic parameter has been developed to help define processes. The new parameter, called Gibbs free energy (G), is defined as being equal to the enthalpy of the system minus the product of entropy and temperature.

Temperature dependence of free energy

Under conditions of constant pressure it is possible to predict the effect on Gibbs free energy of changes in temperature. Following on from the definition of G, it is clear that the change in free energy as a function of temperature is equal to 'negative' entropy. As the change in entropy is always positive, this means that Gibbs free energy always decreases with an increase in temperature.

Pressure dependence of free energy

At constant temperature the change in Gibbs free energy as a function of pressure is equivalent to the volume of the system. As volume is always a positive quantity then Gibbs free energy increases as pressure increases.

Chemical equilibria

It is possible to derive an expression which relates the ideal gas equation to Gibbs free energy, which can then be used to describe equilibrium processes in the gas phase. This may readily be extended to the solution phase to describe chemical reactions.

Related topics

Basic concepts (O1)
First law of thermodynamics (O2)

Second law of thermodynamics (O3)

Gibbs free energy

The equations derived for defining enthalpy and entropy are not convenient in practice, as both the 'system' and the 'surroundings' (which together constitute 'the universe') need to be accounted for. For example, consider a system at thermal equilibrium at a temperature T, in which a process is occurring which involves the release of a small quantity of heat (dQ) to the surroundings (i.e. an **exothermic process**). This means that the heat lost by the system (dQ_{sys}) is absorbed by the surroundings (dQ_{surr}). This small amount of heat does not raise the temperature of the surroundings; in general we always use the simplifying assumption that heat is either taken up or released by the surroundings without causing a temperature change in these surroundings. Because heat lost by the system is gained by the surroundings, it follows that $\Delta H_{surr} = -\Delta H_{sys}$. As we have seen in Topic O3, the entropy change of the surroundings is then $-\Delta H_{sys}/T$.

The total entropy change, dS_{univ}, is the sum of the entropy change of the system (dS_{sys}), and of the surroundings (dS_{surr})

$$\Delta S_{total} = \Delta S_{sys} - \Delta H_{sys}/T$$

For a spontaneous process to occur (see Topic O3), the total entropy cannot decrease, so we can write:

$$\Delta S - \Delta H/T \geq 0$$

where we no longer need the subscripts referring to 'system' and the surroundings'; we can now concentrate on just 'the system'. We can re-arrange this equation to give:

$$\Delta H - T\Delta S \leq 0$$

This equation defines a new function, G, the Gibbs free energy:

$$\Delta G = \Delta H - T\Delta S$$

or

$$G = H - TS$$

This is a convenient measure of the tendency for any chemical process to occur spontaneously, depending on how negative ΔG is. *Table 1* shows the dependence of G on the signs of enthalpy and entropy.

Table 1.   *The effect on ΔG of sign changes in ΔH and ΔS*

ΔH	ΔS	ΔG
+	+	Positive at low temperature, negative at high temperature. Reaction spontaneous in the forward direction at high temperature, and in the reverse at low temperatures.
+	−	Positive at all temperatures. Reactions spontaneous in reverse direction at all temperatures.
−	+	Negative at all temperatures. Spontaneous in forward direction at all temperatures.
−	−	Negative at low temperature, positive at high temperature. Spontaneous at low temperature, reverse at high temperature.

As with enthalpy there is no way to measure absolute values for Gibbs free energies, consequently we define the Gibbs free energy for each of the elements at 1 atm and 298 K to be equal to zero; free energies for molecules under these conditions are referred to as **standard free energies**. Therefore, by analogy to the enthalpy of formation, Gibbs free energy of formation is as follows;

$$\Delta G_f^\circ = \Delta H_f^\circ - T\Delta S_f^\circ$$

There are a number of ways in which the expression for ΔG may be written, and these are useful when considering how temperature and pressure affect this parameter.

In processes involving infinitesimal changes (ΔG becomes dG) then;

$$dG = dH - TdS - SdT$$

for reversible processes;

$$dG = VdP - SdT \qquad \text{(a)}$$

(see Topics O1, O2, and O3).

Temperature dependence of free energy

If ΔG was measured for a reaction at 298 K and it was necessary to know if it would be more favorable at a higher temperature, then of course the reaction could be re-run. Alternatively, we could predict the change in ΔG with temperature at constant pressure, provided the enthalpy for the reaction was known.

Starting with the definition of Gibbs free energy and substituting into equation (a) then;

$$\frac{dG}{dT_p} = -S \quad \text{(this means that as entropy is always positive}$$
$$\text{G always decreases as temperature increases)}$$

It is frequently important to measure G/T as a function of temperature, thus an expression involving enthalpy may be derived;

$$d(G/T)/T_p = -H/T^2 \qquad \qquad \text{(b)}$$

Equation (b) is referred to as the **Gibbs–Helmholtz** equation.

Pressure dependence of free energy

In a similar manner to the temperature dependence of G, it is possible to predict the pressure dependence at constant temperature.

$$dG/dP_T = V$$

As V is always positive, this means that ΔG always increases when pressure increases at constant temperature.

If we consider a pressure change from P_1 to P_2, and utilize the ideal gas equation (see Topic O1), then the following expression may be derived;

$$\Delta G = nRT \ln P_2/P_1$$

If $P_1 = 1$ atm, then

$$G = G° + nRT \ln P \qquad \qquad \text{(c)}$$

where **G°** is the **standard free energy**.

Chemical equilibria

It is possible to relate ΔG for a reaction to the concentration of reacting species and the reaction temperature. To achieve this however reactions in the gas phase need to be considered initially.

Consider the reaction;

$$aA + bB \longrightarrow cC + dD$$

now,

$$\Delta G = G_{prod} - G_{react}$$

using equation (c) $\Delta G°$ may be written as follows;

$$\Delta G = \Delta G° + RT\ln (P_C^c P_D^d/P_A^a P_B^b)$$

where P_A, P_B, P_C, and P_D are the partial pressures of A, B, C, and D.

At equilibrium $\Delta G = 0$, and $(P_C^c P_D^d)/P_A^a P_B^b) = K_p$, the equilibrium constant at constant pressure. Therefore;

$$\Delta G° = - RT \ln K_p$$

We may readily extend this equation to cater for equilibrium processes in the solution phase. In effect we simply replace K_p by K_c, where K_c is given by;

$$K_c = [C]^c[D]^d/[A]^a[B]^b$$

and [] is the concentration in mol per liter.

P1 INTRODUCTION

Key Notes

Rates of reaction

The rate of any chemical reaction is defined as, the change in concentration of one or more components of the reaction over a period of time.

First order reactions

The order of reaction refers to the number of components involved in the rate determining step; that is, the only step of a one step process or the slowest step of a multistep process. Thus, a reaction described as first order depends on the concentration of only one component of the reaction mixture. Order should not be confused with stoichiometry, which is the balance between reactants and products.

Second order reactions

In analogy to first order reactions, second order reactions depend on either the concentrations of two components in the rate determining step, or the concentration squared of one of the components.

Rates for equilibrium and multistep processes

Most chemical reactions are reversible and this needs to be accounted for in determining the overall rate. In multistep processes the rate for each step must be considered and the consequences for the concentrations of intermediates must be taken into account.

Temperature dependence of reaction rates

The rate constants for reactions are independent of reactant or product concentrations but are dependent on reaction temperature, hence so is the reaction rate. In the simplest case, increasing the temperature of a reaction should increase the number of collisions between reacting molecules, and thus increase the rate. However, this is not always the case. Consequently different models for rate versus temperature need to be considered.

Related topics

Determination of reaction order (P2)

Molecularity (P3)

Rates of reaction

The area of chemistry called **chemical kinetics** is concerned with discovering how and why reactions take place. This is achieved by determining **reaction rates** and, by determining the number of reaction steps and the nature of intermediates, the **mechanism of reactions**.

The rate of a reaction is expressed as the change in concentration of reactants (**R**) or products (**P**) over time (**t**). Of course as the concentration of reactant decreases with time, and as the rate can only be positive, a **minus sign** should be included;

$$\text{rate} = \frac{-\Delta[R]}{\Delta t}$$

It is often of interest to determine the **initial rate** of a reaction; thus $[R] = [R]_{initial}$ and $[P]_{initial} = 0$. The formation of products can complicate rate determinations. Therefore, the rate expression should be written as follows;

$$\text{rate} = \frac{-d[R]}{dt} = \frac{d[P]}{dt}$$

When several components are present in a reaction mixture a similar expression may be written. For example;

$$aA + bB \longrightarrow cC + dD \qquad (1)$$

where a, b, c, and d are the **stoichiometries**, that is, the **number of moles** of each component, then the rate is as follows;

$$\text{rate} = \frac{-1d[A]}{a \cdot dt} = \frac{-1d[B]}{b \cdot dt} = \frac{1d[C]}{c \cdot dt} = \frac{1d[D]}{d \cdot dt}$$

First order reactions

The relationship between the rate of a chemical reaction and the concentration of reaction components is complex, and must be determined by experiment. In the case of reaction (1) the rate of the reaction may be written in two forms, one involving the rate of change of concentration, the other simply involving concentration, thus;

$$\text{rate} = \frac{-1d\,[A]}{a \cdot dt}$$

or

$$\text{rate} = k\,[A]^x[B]^y$$

where k is referred to as the **rate constant**, [A] and [B] the concentrations of the reactants, and x and y are referred to as the **orders** such that x + y is the overall **order of reaction**. The order of a reaction should not be confused with stoichiometry, as illustrated by the following example;

$$2\,N_2O_5\,(g) \longrightarrow 4\,NO_2 + O_2$$

By experiment the rate of reaction is found to be;

$$\text{rate} = k\,[N_2O_5]$$

note in this case x = 1, not 2 as stoichiometry would suggest. This and similar processes are therefore referred to as **first order** reactions.

Using the more general reaction (1) the rate for this if it were a first order reaction would be written as;

$$\text{rate} = k'\,[A]^1[B]^0 = k'\,[A]$$

if k = k'a (a is the number of moles of A, and k' the rate constant for conversion of A to products) then;

$$\frac{-d[A]}{dt} = k\,[A]$$

Upon integration;

$$[A] = [A]_0\,e^{-kt}$$

where $[A]_0$ is the initial concentration of A. Consequently a plot of ln $[A]/[A]_0$ versus time will produce a straight line with slope $-k$.

Second order reactions

There are two possibilities for a second order reaction. Consider first reaction (1) under conditions where the rate is found to depend solely on [A], that is;

$$\text{rate} = k' [A]^2 [B]^0$$

and

$$\frac{-1}{a} \frac{d[A]}{dt} = k'[A]^2$$

Integrating produces;

$$\frac{1}{[A]} - \frac{1}{[A]_0} = kt$$

Alternatively if the rate was found to depend on both [A] and [B], that is;

$$\text{rate} = \frac{-1}{a} \frac{d[A]}{dt} = k' [A]^1 [B]^1$$

Several integrations and substitutions lead to the following expression;

$$\frac{1}{[B]_0 - [A]} = \ln \frac{[B][A]_0}{[A][B]_0} = kt$$

where $[A]_0$ and $[B]_0$ are the initial concentrations of A and B.

Rates for equilibrium and multistep processes

Most chemical reactions are reversible to some degree. Clearly then, when considering the rate of a reversible reaction both the forward and backward steps must be considered.

Take the simple example of a reactant A being converted to B, in a reversible manner. There are two elementary steps in such a reaction;

$$A \underset{k_{-1}}{\overset{k_1}{\rightleftharpoons}} B$$

Thus;

$$\frac{-d[A]}{dt} = k_1 [A] - k_{-1} [B]$$

At equilibrium there is no net change in concentration, hence $d[A]/dt = 0$ Therefore;

$$\frac{[B]}{[A]} = \frac{k_1}{k_{-1}} = K = \text{equilibrium constant}$$

When a reaction involves more than one step, then each step of the reaction must be considered when determining a reaction rate.

Consider the following reaction;

$$A \underset{k_{-1}}{\overset{k_1}{\rightleftharpoons}} B \underset{k_{-2}}{\overset{k_2}{\rightleftharpoons}} C$$

$$\frac{-d[A]}{dt} = k_1 [A] \text{ i.e. the reaction is first order and } [A] = [A]_0 e^{-k_1 t}$$

and,

$$\frac{d[B]}{dt} = k_1[A] - k_2[B]$$

and,

$$\frac{d[C]}{dt} = k_2[B]$$

To simplify the determination of k_2 a **steady state approximation** is assumed (see Topic P4) in which the concentration of intermediate B is small but relatively constant. Therefore;

$$\frac{d[B]}{dt} = 0$$

$$= k_1[A] - k_2[B]$$

therefore;

$$[B] = \frac{k_1[A]}{k_2} = \frac{k_1[A]_0\, e^{-k_1 t}}{k_2}$$

As $d[C]/d[B] = k_2 dt$, substituting for [B] and integrating gives;

$$[C] = [A]_0\, (1 - e^{-k_1 t})$$

Temperature dependence of reaction rates

The rate of a reaction would normally be expected to depend on two factors, the number of collisions per second, and the number of those collisions that lead to a reaction. Clearly both of these quantities should increase with temperature, however experiments have shown that this is not always true.

Consider the reaction of nitric oxide molecules in the formation of nitrogen dioxide. The reaction is believed to require two steps; the first of which is an equilibrium process.

$$2NO \rightleftharpoons (NO)_2 \text{ (rapid) } K = \frac{[(NO)_2]}{[NO]^2} \qquad \text{Step 1}$$

$$(NO)_2 + O_2 \xrightarrow{\ k'\ } 2NO_2 \text{ (slow: rate-determining)} \qquad \text{Step 2}$$

The overall rate for this reaction is then;

$$d[NO_2] = k'[(NO)_2][O_2] = k'K[NO]^2\, [O_2]$$

The equilibrium process is exothermic in the forward direction, hence K will decrease with temperature. This outweighs the increase in k′ and therefore increasing the temperature decreases the rate of this reaction.

The example of nitrogen dioxide is not typical but nevertheless serves to illustrate that care must be taken when considering which factors to change to alter the rate of a reaction.

P2 DETERMINATION OF REACTION ORDER

Key Notes

Integration method

Having derived expressions for the rates of chemical reactions, in which orders are specified, an obvious method for determining the order experimentally is to simply measure the concentration of one or more reactant, at specified time intervals during the course of the reaction, and to fit these data to these expressions. This process is referred to as the integration method.

Half-life method

The half-life of a reaction is taken to be the time at which half of the reactants concentration has been used. If the half-life is measured for a series of reactions which differ in the initial concentrations of reactants, then it is possible to derive the reaction order from the previously derived rate expressions.

Differential method

If the rate of a reaction is measured for a series of concentrations of reactants, then it is possible to determine the reaction order as this is simply the slope of a plot of log (rate) versus log (concentration). This process is referred to as the differential method for determining reaction order.

Related topics   Introduction (P1)   Molecularity (P3)

Integration method

There are a number of approaches which may be adopted in order to determine the **order** of a reaction. The first of these, known as the **integration method**, is perhaps the most obvious. This simply depends on the measurement of concentrations of reactants (one or all could be followed) at specific times during the course of the reaction. Once this is done the data obtained is simply substituted into the derived equations for reaction rates (see *Table 1*). The equation giving the most constant value for the rate constant over a series of time intervals is the one corresponding to the correct order for the reaction.

Table 1. *Summary of the rate equations derived for reactant (A) going to products*

Order	Differential form	Integrated form	Half-life	Units of rate constant (k)
0	$\dfrac{-d[A]}{dt} = k$	$[A]_0 - [A] = kt$	$\dfrac{[A]_0}{2k}$	$M\ s^{-1}$
1	$\dfrac{-d[A]}{dt} = k[A]$	$[A] = [A]_0\ e^{-kt}$	$\dfrac{\ln 2}{k}$	s^{-1}
2	$\dfrac{-d[A]}{dt} = k[A]^2$	$\dfrac{1}{[A]} - \dfrac{1}{[A]_0} = kt$	$\dfrac{1}{[A]_0 k}$	$M^{-1}s^{-1}$

Half-life method   The **half-life** ($t_{1/2}$) of a reaction is the time taken for half of the reactants to be used up. Therefore, for a **first order** reaction (see Topic N1) the following expression may be written;

$$[A] = [A]_0/2 \text{ at time } t = t_{1/2} \text{ (half way through the reaction)}$$

$$\text{hence } \ln \frac{[A]_0/2}{[A]_0} = -kt_{1/2}$$

and therefore

$$t_{1/2} = \ln 2/k = 0.693/k$$

To utilize this approach the dependence of the half-life on the initial concentration of reactants must be determined. This is achieved by following a series of reactions which differ in the initial concentrations of reactants. It is then possible to derive the reaction order from the previously derived rate expressions. In general a reaction of the nth order (see *Table 1*) has a half life given by;

$$t_{1/2} = \frac{1}{k[A]_0^{n-1}}$$

Differential   The **differential method** was first suggested by **van't Hoff** towards the end of the
method   19th century and depends on performing experiments with different concentrations of reactants and measuring the rate for each experiment.

As the rate of an nth order reaction is dependent on the nth power of the concentration of a reactant, we can write;

$$\text{rate} = v = k[A]^n$$

and therefore

$$\log v = n \log [A] + \log k$$

Thus, a plot of [A] versus time, for a series of A concentrations, enables the initial rate, v_0, to be determined at each concentration. Thereafter a plot of $\log v_0$ versus $\log [A]$ enables n to be determined.

P3 MOLECULARITY

Key Notes

Definition of molecularity	The term molecularity of a reaction defines the number of species involved in an elementary step of a reaction (the only step in a single step reaction). The molecularity of a reaction may not be immediately apparent from the stoichiometry of a reaction. When the molecularity of each step of a reaction is known, then the mechanism for the reaction may be elucidated.
Unimolecular reactions	A unimolecular reaction is a reaction in which only one type of molecule is involved in producing an intermediate or a reaction product. To rationalize this we have to introduce the possibility of molecules becoming activated by collisions with like molecules; these activated molecules then having enough energy to undergo a change to form the reaction product.
Related topic	Introduction (P1)

Definition of molecularity

It is very rarely the case that a reaction takes place in the manner suggested by the overall chemical equation for that reaction. It is generally the case that in converting reactants to products, a number of **elementary steps** are required; involving bond making and breaking, the formation of stabilizing interactions, the formation of intermediates, and so on. All of these steps must be known for the mechanism of the reaction to be understood. The mechanism of any reaction must be able to account for the overall stoichiometry and the rates determined for each of the steps involved.

To illustrate the difference between **molecularity** and stoichiometry consider the conversion of hydrogen peroxide to oxygen and water, a reaction which is **catalyzed** (that is accelerated) by the presence of iodide ions (see *Fig. 1*).

As both reaction (1) and (2) involve two reagents, each reaction is said to have a **molecularity** of 2, or to be **bimolecular**. To rationalize the rate expression on the basis of a two step mechanism, it has to be assumed that the rate of step (1) is very

$$2H_2O_2 \xrightarrow{k} 2H_2O + O_2$$

$$\text{Rate} = \frac{-d[H_2O_2]}{dt} = k[H_2O_2][I^-]$$

The reaction appears to be second order, depending equally on the concentration of peroxide and iodide ions.
The actual mechanism for the reaction is thought to involve two elementary steps;

(1) $H_2O_2 + I^- \longrightarrow H_2O + IO^-$
(2) $H_2O_2 + IO^- \longrightarrow H_2O + O_2 + I^-$

Fig. 1. The iodide catalyzed conversion of hydrogen peroxide to oxygen and water.

much slower than the rate of step (2); therefore step (1) is the **rate determining step**. This is a trivial example, but serves to illustrate that complete understanding of chemical reactions comes from a knowledge of molecularities, not from reaction orders.

Unimolecular reactions

Unimolecular reactions are those which involve only one type of molecule in an elementary step. This is quite a difficult concept to grasp. To help rationalize observations that such reactions do indeed occur; examples of such processes include **thermal decompositions**, *cis-trans* **isomerizations**, and so on. Lindermann, in 1922, proposed that a reactant molecule A, must collide with another like molecule. After the collision one of the the molecules, A*, is more excited, at the expense of the other A, and has sufficient energy to convert into a product (see *Fig. 2*).

$$(1) \quad A + A \xrightarrow{k_1} A + A^*$$

$$(2) \quad A^* \xrightarrow{k_2} Product$$

The possibility of the reverse of reaction (1) occurring must also be considered

$$A^* + A \xrightarrow{k_{-1}} A + A$$

Fig. 2.   A proposed mechanism for unimolecular processes.

The rate for the overall reaction may be written in terms of the reaction product;

$$\frac{d[product]}{dt} = k_2\,[A^*]$$

However, as A* is an excited molecule it will be unstable and therefore its concentration will be both low and relatively constant. Therefore, it is reasonable to utilize **a steady state approximation**; that is, to assume the rate of change in concentration of A* is zero. Thus;

$$\frac{d[A^*]}{dt} = 0$$

and,

$$\frac{d[product]}{dt} = \frac{k_1 k_2 [A]^2}{k_2 + k_{-1}[A]}$$

This expression may be simplified when considering two extreme situations.

At pressures of 1 atm and above it is likely that A* will be deactivated due to collisions with A more frequently than it can form products, hence;

$$k_{-1}\,[A][A^*] \gg k_2\,[A^*] \text{ or } k_{-1}[A] \gg k_2$$

therefore,

$$\frac{d[product]}{dt} = \frac{k_1 k_2 [A]}{k_{-1}}$$

thus, the **unimolecular** reaction is also **first order**.

In contrast at lower pressures, collisions between A* and A become less likely and hence;

$$k_{-1}[A][A^*] \ll k_2[A^*] \text{ or } k_{-1}[A] \ll k_2$$

therefore

$$\frac{d[\text{product}]}{dt} = k_1[A]^2$$

the **unimolecular** reaction is **second order** under these conditions.

Unimolecular processes although known are rather less common than those involving two species, that is, bimolecular processes; numerous examples of which appear throughout this text. It should be noted, however, that **termolecular** (involving three species) and higher molecularity reactions are relatively rare due to the low probability of three or more bodies (molecules) being able to collide with each other at the same time.

P4 ENZYME KINETICS

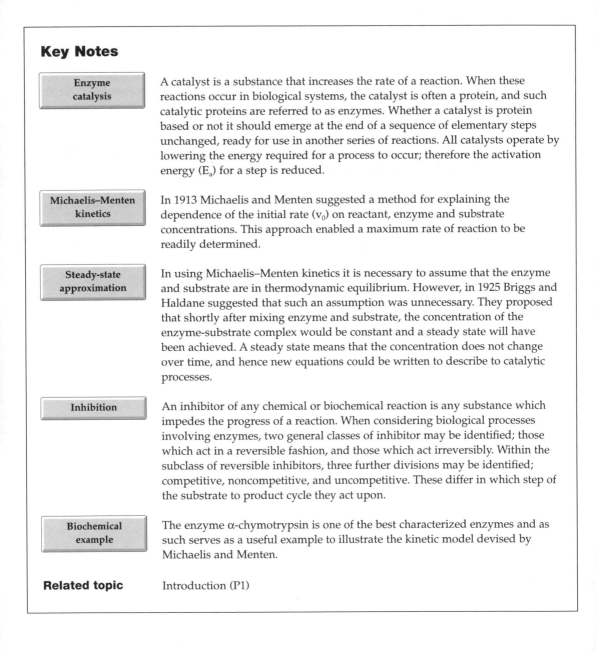

Key Notes

Enzyme catalysis

A catalyst is a substance that increases the rate of a reaction. When these reactions occur in biological systems, the catalyst is often a protein, and such catalytic proteins are referred to as enzymes. Whether a catalyst is protein based or not it should emerge at the end of a sequence of elementary steps unchanged, ready for use in another series of reactions. All catalysts operate by lowering the energy required for a process to occur; therefore the activation energy (E_a) for a step is reduced.

Michaelis–Menten kinetics

In 1913 Michaelis and Menten suggested a method for explaining the dependence of the initial rate (v_0) on reactant, enzyme and substrate concentrations. This approach enabled a maximum rate of reaction to be readily determined.

Steady-state approximation

In using Michaelis–Menten kinetics it is necessary to assume that the enzyme and substrate are in thermodynamic equilibrium. However, in 1925 Briggs and Haldane suggested that such an assumption was unnecessary. They proposed that shortly after mixing enzyme and substrate, the concentration of the enzyme-substrate complex would be constant and a steady state will have been achieved. A steady state means that the concentration does not change over time, and hence new equations could be written to describe to catalytic processes.

Inhibition

An inhibitor of any chemical or biochemical reaction is any substance which impedes the progress of a reaction. When considering biological processes involving enzymes, two general classes of inhibitor may be identified; those which act in a reversible fashion, and those which act irreversibly. Within the subclass of reversible inhibitors, three further divisions may be identified; competitive, noncompetitive, and uncompetitive. These differ in which step of the substrate to product cycle they act upon.

Biochemical example

The enzyme α-chymotrypsin is one of the best characterized enzymes and as such serves as a useful example to illustrate the kinetic model devised by Michaelis and Menten.

Related topic

Introduction (P1)

Enzyme catalysis

The definition of a catalyst is, a substance which increases the rate of a reaction. The catalyst may change during the course of a reaction but is regenerated in its original form at the end of the process. Irrespective of the type of process or mechanism of reaction involved a catalyst operates by lowering the energy required; this is the activation energy (E_a), for the rate determining step. When considering biological processes the majority of catalysts take the form of proteins; proteins acting in this capacity are called **enzymes**.

Enzymes differ in general from chemical or small molecule catalysts in the degree of specificity they display. Often an enzyme will only catalyze the conversion of one particular compound (or class of compound), referred to as the **substrate**, to product. This specificity is achieved by the 3–dimensional arrangement and types of functional group at the **active site** of the enzyme. This is the region of the protein where chemical processes occur. Towards the end of the 19th century Fischer described the **lock and key** model for enzyme recognition of substrates. In this the active site of the enzyme is seen as the lock, and the substrate as having the correct structure and functional group characteristics to act as a key. This model takes no account of flexibility of either active site or substrate, now known to be an important factor, but nevertheless is a useful descriptive model.

When a catalyst is involved in a reaction, clearly it must be accounted for in rate expressions. In deriving modified expressions it is customary to consider only initial rates (v_0). As a reaction progresses it is possible that the efficiency of the enzyme may be inhibited by the product, by pH changes and so on.

If we assume that enzyme (E) and substrate (S) form a 1 : 1 complex (ES) then the initial rate should increase up to the point when the enzyme is **saturated** or fully complexed, where the rate levels off (see *Fig. 1*). The relationship between substrate concentration ([S]) and initial rate, is a hyperbolic function.

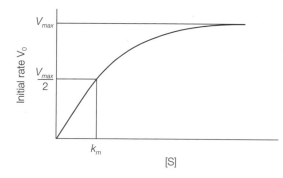

Fig. 1. Michaelis–Menten plot of initial rate of an enzyme catalyzed reaction versus substrate concentration. V_{max}, maximum rate; K_m, Michaelis constant, at fixed enzyme concentration.

Michaelis–Menten kinetics

In 1913, Michaelis and Menten devised a model to explain the dependence of the initial rate on substrate concentration. Consider the basic reaction scheme;

$$E + S \underset{k_{-1}}{\overset{k_1}{\rightleftharpoons}} ES \xrightarrow{k_2} P + E$$

where P is the product.

Thus, the initial rate of product formation is then;

$$v_0 = \frac{d[P]}{dt_0} = k_2 [ES] \tag{1}$$

If we assume that at time zero the substrate concentration is much greater than the enzyme concentration, and also therefore much greater than the complex concentration, then it follows that [S] is approximately equal to $[S]_0$. If we further assume that the rate constant for the reverse reaction, k_{-1} is much greater than k_2, then we can write an expression for the substrate dissociation constant, K_s as follows;

$$K_s = \frac{k_{-1}}{k_1} = \frac{[E][S]}{[ES]} \qquad (2)$$

The concentration of E at any time is clearly $[E]_0 - [ES]$, and substituting for this in expression (2) and subsequently for [ES] in equation (1) we arrive at the following;

$$v_0 = \frac{k_2[E]_0[S]}{K_s + [S]} \qquad (3)$$

where $k_2[E]_0$ and K_s can be thought of as the constants.

At high substrate concentrations;

$$v_0 \approx k_2[E]_0 \qquad (4)$$

the rate is constant. At this point we refer to a **maximum rate** symbolized V_{max}.

Steady-state approximation

The first step of the Michaelis–Menten model of enzyme kinetics makes the assumption that the enzyme and the substrate are in thermal equilibrium with the enzyme-substrate complex. In the early 1900s Briggs and Haldane proposed a model which did not require this assumption. The proviso for the new model was that the ES complex be formed shortly after mixing the two components and that the concentration of complex was then relatively constant; that is, a **steady state** is set up (see *Fig. 2*).

Therefore, assuming that d[ES]/dt is zero it is possible to derive the following expression;

$$v_0 = \frac{V_{max}[S]}{K_m + [S]} \qquad (5)$$

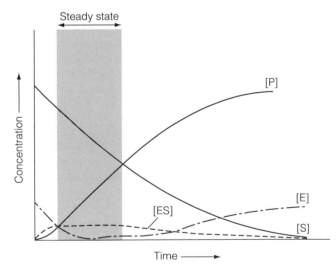

Fig. 2.   *Plot of concentrations of components of an enzyme catalyzed reaction versus time. ES, enzyme substrate complex concentration; E, enzyme concentration; S, substrate concentration; P, product concentration.*

where K_m is the **Michaelis constant** and is equal to $(k_{-1} + k_2)/k_1$.
As the initial rate is half the maximum rate (5) simplifies to;

$$K_m = [S] \qquad (6)$$

In practice it is quite difficult to obtain V_{max} from a plot of v_0 versus $[S]$, particularly at high substrate concentrations. Consequently, **Lineweaver and Burk** suggested employing double reciprocals. Thus;

$$1/v_0 = (K_m/V_{max}[S]) + 1/V_{max}$$

a plot of which is shown in *Fig. 3*.

It is worth noting that k_2 is commonly referred to as the **turnover number**; that is, the number of substrate molecules converted to product per enzyme molecule per unit of time. It is also known as k_{cat}. Also, K_m may be interpreted in terms of the strength of the enzyme-substrated complex; large values of K_m for a weakly bound complex, small values for a tightly bound system.

Inhibition

Any substance which decreases the rate of a reaction is referred to as an **inhibitor**. Although this would seem a rather negative property, enzyme inhibitors are frequently sought in order to elucidate active site mechanisms and to probe functional groups at the active site.

There are two general classes of enzyme inhibitors. Inhibitors labeled **reversible** are involved in an equilibrium with the enzyme or enzyme substrate complex. Those referred to as **irreversible** inhibit the enzyme progressively, eventually when the concentration of the inhibitor exceeds that of the enzyme, inhibition is complete. Within the class of reversible inhibitors there are three further subdivisions; **competitive**, **noncompetitive**, and **uncompetitive** inhibitors. If an inhibitor (I) is **competitive** it will bind to the same site as the substrate;

$$E + S \longrightarrow ES \longrightarrow E + P$$
$$E + I \longrightarrow EI \longrightarrow no\ product$$

If we assume Michaelis–Menten kinetics and produce a Lineweaver–Burk plot we find that at constant inhibitor concentration, the intercept of the line is as for the uninhibited situation, but the slope is enhanced by a factor dependent on $[I]$.

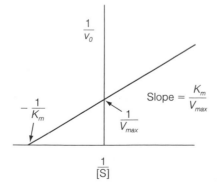

Fig. 3. Lineweaver–Burk plot for enzyme catalyzed reactions following Michaelis–Menten kinetics. K_m, Michaelis constant; V_o, initial rate; $[S]$, substrate concentration; V_{max}, maximum rate.

The effects of a competitive inhibitor can readily be overcome by simply increasing [S]. If an inhibitor is **noncompetitive** then it generally does not bind to the active site;

$$E + S \longrightarrow ES \longrightarrow E + P$$
$$E + I \longrightarrow EI \longrightarrow \text{no product}$$
$$ES + I \longrightarrow ESI \longrightarrow \text{no product}$$

As the inhibitor does not interfere with the substrate binding site, its effects cannot be reversed by increasing the substrate concentration.

If an inhibitor is **uncompetitive** then it does not bind to the free enzyme, only to the enzyme substrate complex;

$$E + S \longrightarrow ES \longrightarrow E + P$$
$$ES + I \longrightarrow ESI \longrightarrow \text{no product}$$

Once again its effects cannot be reversed by simply increasing the concentration of S. In *Fig. 4* the Lineweaver–Burk plots for all three types of reversible inhibition are shown.

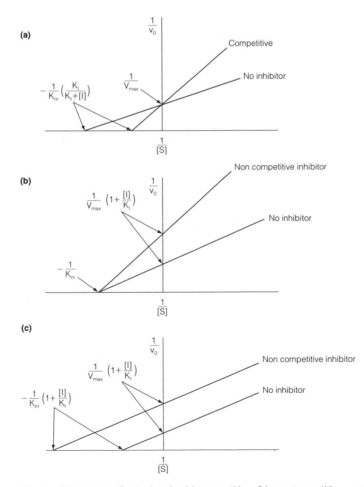

Fig. 4. Lineweaver–Burk plots for (a) competitive, (b) noncompetitive, and (c) uncompetitive inhibition. V_{max}, maximum rate; [S], substrate concentration; K_m, Michaelis constant; V_0, initial rate; [I], inhibitor concentration.

Biochemical example

The classic example for illustrating enzyme kinetics is α-chymotrypsin, which is one of the most studied and best understood enzymes. It is a member of a class of enzymes referred to as the serine proteases; these are all proteins which have a serine residue in their active site and their function is to cut other proteins at specific sites.

One of the reactions catalyzed by α-chymotrypsin is the hydrolysis of p-nitrophenyltrimethylacetate (a) to p-nitrophenol (b) (see *Fig. 5*)

As both (a) and (b) are uv/vis active (see Topic Q2) but with different λ_{max} (nitrophenol has a strong (visible) color) the reaction may be followed using this spectrophotometric technique. Experiments have shown that the enzyme catalyzed reaction involves two rate steps;

$$E + S \underset{k_{-1}}{\overset{k_1}{\rightleftharpoons}} ES \overset{k_2}{\longrightarrow} ES' + P_1 \overset{k_3}{\longrightarrow} E + P_2$$

where P_1 is nitrophenol, P_2 is trimethylacetate, and ES' is referred to as the trimethylacetyl enzyme as at this stage the trimethylacetyl group is covalently linked to the serine residue in the active site (see *Fig. 5*). This covalent bond must be broken so that the enzyme may be involved in another catalytic cycle (see *Fig. 5*). The catalytic rate constant for this sequence, k_{cat} is given by;

$$k_{cat} = (k_2 k_3)/(k_2 + k_3)$$

In *Table 1* the rate constants for the reactions referred to are provided. These were obtained by fitting the various rate equations to the curve obtained by plotting the absorbance at 400 nm (see Topic Q2) of p-nitrophenol against time.

Table 1.   Values for some of the rate variables determined from the kinetic analysis of the α-chymotrypsin catalyzed hydrolysis of p-nitrophenyltrimethylacetate (at pH 8.2)

Variable	Value	Constant
k_2	$0.37 +/- 0.11 \text{ s}^{-1}$	rate constant for ES → P + E (turnover number)
K_s	$1.6 +/- 0.5 \times 10^{-3} \text{ M}$	substrate dissociation constant
k_{cat}	$1.3 \times 10^{-4} \text{ s}^{-1}$	rate constant for catalysis
K_m	$5.6 \times 10^{-7} \text{ M}$	Michaelis constant

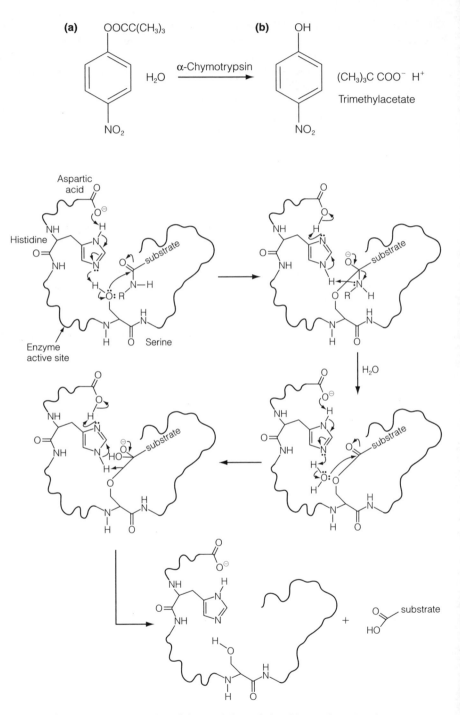

Fig. 5.   *Schematic representation of the catalytic cycle involving α-chymotrypsin.*

P5 CATALYSIS AND THERMODYNAMIC VERSUS KINETIC CONTROL

Key Notes

Activation energy

A chemical reaction which is thermodynamically favorable may not proceed at an observable rate if there is a significant energy barrier lying between the reactants and the products. This is sometimes called a kinetic barrier. The energy required to surmount this barrier is called the activation energy. Increasing the temperature will increase the reaction rate as the barrier becomes easier to cross. However, if the barrier is particularly high the reaction will not proceed at a significant rate, even though the reaction is thermodynamically favored.

Catalysis

A catalyst reduces the height of the kinetic barrier, and thus accelerates the reaction, but does not change the relative energies (stabilities) of the reactants or products, and so does not alter the position of equilibrium. Catalysts in biological systems are generally specialized proteins called enzymes. These are able to catalyze reactions in a highly specific manner, often to very fast rates.

Thermodynamic and kinetic control

A reaction for which the kinetic barrier is not significant, such as one which is efficiently enzyme-catalyzed, is said to be under thermodynamic control. This is because thermodynamic equilibrium is achieved, and thus only thermodynamic considerations are needed to assess the relative amounts of reactants and products. Conversely, if the kinetic barrier is so high it cannot be crossed, thermodynamics do not define the system which is then said to be under kinetic control.

Related topics

Gibbs free energy (O4)   Enzyme kinetics (P4)

Activation energy

A chemical reaction can be represented in the form shown in *Fig. 1*.

Here the products are of lower energy than the reactants, so the reaction is thermodynamically favored, and thus is possible in principle. However, for the reactant to become the product, an energy barrier must be surmounted, which is sometimes referred to in general terms as a **kinetic barrier**. The height of this, with respect to the energy of the reactant, represents the **activation energy** of the reaction; the energy that the reactant must possess in order to be able climb over this barrier. At a given temperature only a small proportion of reactant molecules may possess this much energy at any instant, so these are able to proceed with the reaction. If the temperature is raised, this proportion increases, and so more molecules are able to react and the observed reaction rate is increased. This is why in the laboratory, heating is often used to accelerate chemical reactions. However, if the kinetic barrier is especially high, the reaction will not proceed at any appreciable rate. Thus, although the product is more stable than the reactant, it may not form.

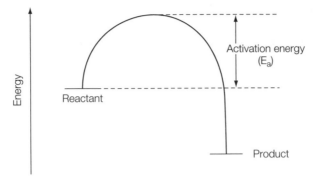

Fig. 1.   Schematic representation of the energy barrier to be crossed between reactants and products.

Thus, although paper is less stable than its combustion products, books do not tend to spontaneously ignite!

Catalysis   A catalyst increases the rate of the reaction by reducing the activation energy (see *Fig. 2*).

It is important to note that a catalyst does not alter the relative energies of the reactants and products, and so does not change the equilibrium state. In other words, a **catalyst increases the rate** at which equilibrium is reached, but **does not alter the position** of that equilibrium.

Most chemical reactions occuring within a biological system are catalyzed by enzymes (see Topic P4). These have evolved to become very efficient at the tasks they are required to perform, both in terms of the specificity of a reaction, and the rate at which it is catalyzed. As the catalytic abilities of an enzyme are generally consequences of its specific protein structure, the enzyme will not function if that structure is perturbed. Thus, as the temperature is raised, the enzyme activity (as manifested in its turnover number, see Topic P4) will increase, since the kinetic barrier although reduced, is still of finite height. Then, at some point the protein itself will begin to thermally denature, and the active site of the enzyme will deform so it no longer functions efficiently. The activity thus drops off. In contrast to an uncatalyzed reaction, enzyme catalysis generally displays an optimum temperature (see *Fig. 3*). For many enzymes this is around $37\,^{\circ}\mathrm{C}$; body temperature.

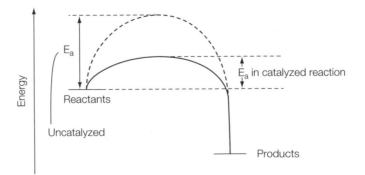

Fig. 2.   The effect on a catalyst of the energy of activation of a reaction.

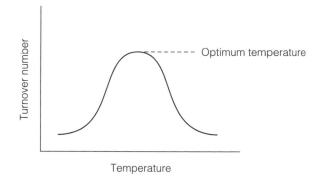

Fig. 3.   Schematic representation of the efficiency of an enzymic catalyst as a function of temperature.

In general, chemical reactions which are not physiologically advantageous will not be catalyzed by enzymes, as there would be no reason for such enzymes to have evolved. Such reactions may include changes (i.e. damage) to functional groups in proteins and nucleic acids, for example. As such processes are not actively catalyzed, the activation energies can be relatively high, and so the reactions may proceed relatively slowly.

Thermodynamic and kinetic control

When equilibrium has been achieved, the composition of a solution of various chemicals will be defined by thermodynamics. This means that we can use thermodynamic calculations to determine the relative proportions of all the chemical species that will be present. The system can then be said to be **under thermodynamic control**. However, if one or more kinetic barriers are prohibitively high (which may be the case if there are no enzymes to catalyze the relevant reactions) then the reaction may not proceed at all, or may occur at only a very slow rate. The composition of the system then depends on kinetic considerations rather than thermodynamics, and so we have **kinetic control**. Achieving a stable balance between thermodynamic and kinetic control is one of the most important features of living organisms.

Q1 THE ELECTROMAGNETIC SPECTRUM

Key Notes

Quantization of energy	Until the end of the 19th century, the absorption or emission of radiation could be explained by classical Newtonian mechanics. However, as more experimental data became available it became clear that absorption and emission were not continuous processes, but that rather they involved discrete packets of energy. This quantization of energy could not be explained until early in the 20th century.
Regions of the spectrum	The electromagnetic spectrum, as the name suggests, has both electric and magnetic components, and covers a huge energy range. The spectrum is generally divided into regions referred to on the basis of the processes which may be detected by the application of radiation of the particular wavelength. The boundaries between these regions are not precise.
Basic instrumentation	Although the details of the instruments which utilize the various regions of the spectrum differ considerably, there are common features relating to them. In general two types of instrument may be considered, one for monitoring absorbed radiation, and one for monitoring emitted radiation. Absorption instruments often have 'grating devices' as an integral part, and of course a light source. Emission instruments differ in that after a sample has been excited it then acts as its own light source.
Related topics	Ultraviolet-visible spectrophotometry (Q2) Circular dichroism (Q5) Fluorescence (Q3) Nuclear magnetic resonance spectroscopy (Q6) Infrared spectrophotometry (Q4)

Quantization of energy

The term spectroscopy refers to an **experimental** subject which is concerned with the **absorption**, **emission**, or **scattering** of electromagnetic radiation. The earliest spectroscopic observations were made using the most accessible region of the electromagnetic spectrum; that is, visible light. In 1665 Newton started experiments on the dispersion of white light in to a range of colors using a glass prism. Over two centuries later Balmer fitted the visible spectrum of atomic hydrogen to a mathematical formula. In doing so he showed that **discrete** wavelengths had to be considered. The classical Newtonian picture would require the spectrum to show a **continuous** change with wavelength and hence was at odds with the new observations. A detailed explanation of the hydrogen spectrum was not possible before the work of Planck at the start of the 20th century. He proposed that tiny **oscillators**; atoms or molecules for example, have an oscillation frequency υ which is related to the energy of emitted radiation E by the expression;

$$E = nh\upsilon$$

where n is an integer and h is known as **Planck's constant**.

The energy therefore is said to be **quantized** in discrete packets, or **quanta**, each of energy $h\upsilon$.

If this is applied for example, to the rotational properties of a molecule, then two rotational states will have energies E_1 and E_2 (see *Fig. 1*); the suffixes 1 and 2 are in fact **quantum numbers**. To move from one state to the other an appropriate amount of energy, ΔE, must be absorbed or emitted, where ΔE is given by;

$$\Delta E = E_2 - E_1$$
$$= h\upsilon \text{ (joules)}$$

In talking about transitions between energy levels it is most usual to refer to frequency, wavelength or wavenumber terms (see Topic Q4) as if these were energy units. Thus, in referring to an 'energy of 100 cm^{-1}' what is actually meant is a 'separation between two energy states, such that the associated radiation has a wavenumber value of 100 cm^{-1}'. It should be remembered that the frequency of radiation associated with an energy change does not imply that the transition between energy states occurs a number of times per second!

Fig. 1.   *Diagram representing two successive rotational energy levels for a molecule.*

Regions of the spectrum

Electromagnetic radiation, as the name implies, contains both an electric and magnetic component. The electric component of **plane polarized light** is in the form of an oscillating electric field, the magnetic component an oscillating magnetic field. These oscillations are sinusoidal and at right angles to each other. The plane of polarization is by convention taken to be the plane containing the direction of the electric component. The reason for this choice is that interaction of electromagnetic radiation with matter, is most commonly through the electric component.

The full energy range of the electromagnetic (EM) spectrum is usually divided up into regions corresponding to the processes which may be monitored by the absorption or emission of energy in that region (see *Fig. 2*). At the high energy (or high frequency) end of the spectrum we are dealing with γ-rays, at the low energy end radio waves are found. The boundaries between these regions are not precise.

A detailed description of the uses of the various regions of the EM spectrum is provided in the remainder of this section.

Basic instrumentation

The instrumentation utilized for the spectroscopic techniques described in Topics Q2–Q6, differ considerably in their detail. However, in general they may be divided into two types; **absorption instruments** and **emission instruments**.

The basic type of absorption instrument, which may be used for ultra-violet, visible and infrared measurements is shown in schematic form in *Fig. 3*. The main components of the instrument are (i) a **source**, which is generally white light, then a guiding device such as a mirror to ensure that the light reaches the sample, (ii) a **sample cell**, which contains windows made of material that permits the light to be transmitted, (iii) an **analyzer**, which is usually a dispersion grating which

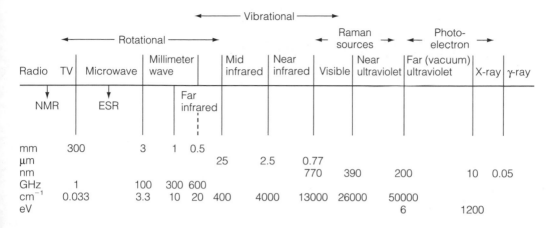

Fig. 2.   The regions of the electromagnetic spectrum. Reprinted, with permission, from Modern Spectroscopy, Hollas, copyright John Wiley and Sons Limited.

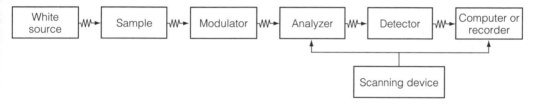

Fig. 3.   Schematic representation of an absorption spectrometer. Reproduced from, C.N. Banwell and E.M. McCash Fundamentals of Molecular Spectroscopy 4th edn. 1994 with the kind permission of McGraw-Hill Publishing Company.

selects the frequency reaching the detector at any given time, and (iv) a **detector**, the recorder of which is synchronized with the analyzer so as to produce a trace of absorbance as the frequency varies.

The layout for an emission instrument (see *Fig. 4*) differs in that after excitation; which may be achieved by heating the sample, passing a large electric current through it, or exposing it to EM radiation, it acts as its own light source. It is therefore only necessary to **collect**, **analyze**, and **record** the emitted radiation. If EM radiation is used to excite the sample then it is important to ensure that this radiation is not detected with the emission spectrum. A **modulator**; a device which interrupts the radiation beam at regular intervals, when coupled to a (frequency) tuned **detector-amplifier**, ensures that the emission recorded from the sample arises directly from the excitation and not the excitation source.

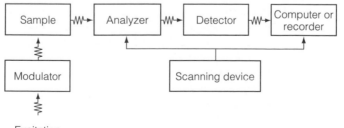

Fig. 4.   Schematic representation of an emission spectrometer. Reproduced from, C.N. Banwell and E.M. McCash Fundamentals of Molecular Spectroscopy 4th edn. 1994 with the kind permission of McGraw-Hill Publishing Company.

Q2 ULTRAVIOLET-VISIBLE SPECTROPHOTOMETRY

Key Notes

Electronic transitions	The energy associated with the ultraviolet and visible regions of the electromagnetic spectrum is sufficient to promote an electron from one energy level to a higher level.
Beer–Lambert law	The Beer–Lambert law relates the amount of UV/visible radiation absorbed to the concentration of the absorbing sample.
Spectral representation	The result of an experiment is a plot of absorbance against wavelength. UV/vis spectra typically display broad lines.
Chromophore	The region of a molecule giving rise to the dominant UV/vis response is called a chromophore.
Uses in biology	UV/vis spectrophotometry is used to determine sample concentrations, to monitor ligand binding, and to monitor conformation changes in biomolecules.
Related topics	The electromagnetic spectrum (Q1) Fluorescence (Q3)

Electronic transitions

In UV/vis spectroscopy, the transitions which result in the absorption of electromagnetic radiation are electronic transitions. An electron is promoted from the highest occupied molecular orbital (HOMO) to the lowest unoccupied molecular orbital (LUMO) (see Topic B2).

In many saturated molecules the HOMO is a sigma (σ) orbital (see Topic B1). Pi (π) orbitals are of higher energy and those occupied by lone-pair (unshared) electrons (nonbonding orbitals, n) at even higher energy (see *Fig. 1*).

Not all transitions are equally probable; selection rules exist. If a transition is **forbidden** this means that it will occur with a **low probability** and will therefore lead to a low intensity band in the spectrum. The n—π^* transition is an important example of a formally forbidden transition.

UV/vis spectrophotometry may be used to monitor the coordination of ligands to heavy metal ions. In these instances, it is transitions amongst d-orbitals that are of interest (see *Fig. 2*). All d—d transitions are forbidden and therefore give a weak UV/vis response, but nevertheless may be strong enough to produce a noticeable color.

Beer–Lambert law

The intensity of light passing through a sample falls off exponentially as it progresses through that sample. The experimental measure of the intensity of

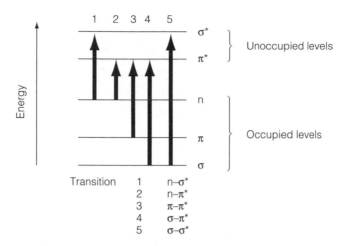

Fig. 1.   Some of the possible electronic transitions.

emerging light, **at a particular wavelength**, is referred to as the **extinction coefficient**, ε, and this is given by the Beer–Lambert law;

$$\log_{10}\left(\frac{I_0}{I_t}\right) = \varepsilon c l$$

where I_0 is the intensity of incident light, I_t the intensity of transmitted light, c is the concentration, l is the length of the path over which light travels. As the difference in I_0 and I_t is a measure of the radiation absorbed the Beer–Lambert law is often expressed as;

$$A = \varepsilon c l$$

where A is the **absorbance** or **optical density**, and is dimensionless. Practically in biology, c is in mol l^{-1}, l is in cm, and ε is the absorbance of a 1 mol l^{-1} solution, with 1 cm path length.

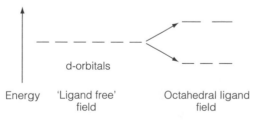

Energy   'Ligand free'   Octahedral ligand
           field   field

Fig. 2.   The degeneracy of d-orbitals depends on coordinated ligands and affects the appearance of the UV/vis spectra of such compounds.

Spectral representation

The UV/vis response of a molecule is generally displayed as a trace of absorbance against wavelength (see *Fig. 3*).

The absorbance maxima are converted to extinction coefficient maxima, ε_{max}, these and the corresponding wavelength maxima, λ_{max}, are reported.

The spectra for the most part are broad and relatively featureless. The broad appearance is due to the fact that molecules are in a constant state of vibration, and consequently vibrational transitions are superimposed with the electronic transitions. This therefore means that in effect each electronic transition consists of

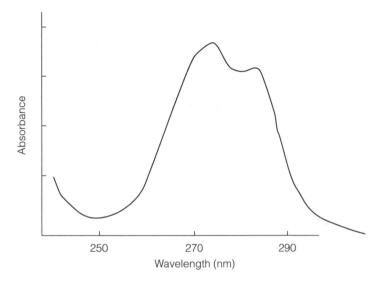

Fig. 3.   UV/vis spectra are generally broad as illustrated by the spectrum of the amino acid tryptophan in water solution.

a vast number of lines, so closely spaced that they can not be resolved by the UV/vis detector. Rather an **absorption band** is detected (see *Fig. 4*).

The appearance of the spectrum can be affected by the nature of the solvent used. A nonpolar solvent does not hydrogen bond with the solute and therefore the spectrum is similar to that in the gas phase, i.e. it has **fine structure**. As a polar solvent does not form hydrogen bonds so readily with the excited state system, the λ_{max} moves to shorter wavelength (see *Table 1*).

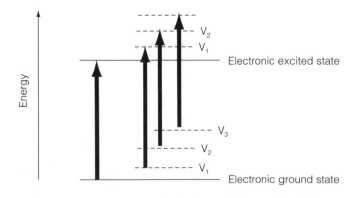

Fig. 4.   The broad appearance of UV/vis spectra may be explained in part by consideration of the vibrational levels which accompany each electronic state.

Table 1.   Effect of solvent on the n—π* transition of acetone

Solvent	H_2O	CH_3OH	$CHCl_3$
λ_{max}(nm)	264.5	270	272

◄─────────────────────────────── Polarity

Chromophore

Chromophore is the term given to a group of atoms giving rise to a strong electronic response. Although electron transitions are being monitored, the degree to which nuclei attached in bonds hold on to their electrons determines the energy of each electronic transition and therefore the wavelength of absorption. In the region of the spectrum most readily accessible, groups of atoms connected by pi bonds are responsible for the most noticeable absorptions. Systems in which pi bonds are conjugated (see Topic J1) give a very strong response in the middle of the UV range (see *Table 2*).

The addition of atoms or groups of atoms to a chromophore can effect the λ_{max} of the chromophore, if the substituent has lone-pairs of electrons which can interact with the pi system. Such substituents are referred to as **auxochromes**. If the auxochrome shifts the λ_{max} to longer wavelengths, then a red or **bathochromic** shift is said to have occurred. A shift to shorter wavelength is referred to as a blue or **hypsochromic** shift. Often these shifts can be accompanied by a change in the absorbance of the molecule. If the absorbance decreases this is referred as a change in **hyperchromicity**, the converse is a change in **hypochromicity**.

Table 2.   Absorptions of isolated chromophores

Chromophore		Transition	λ_{max}(nm)	$\log_{10} \varepsilon$
Alcohol	R – OH	n — σ*	180	2.5
Amine	R – NH₂	n — σ*	190	3.5
Thiol	R – SH	n — σ*	210	3.0
Aldehyde	R – CHO	π — π*	190	2.0
		n — π*	290	1.0
Ketone	R₂ – CO	π — π*	180	3.0
		n — π*	280	1.5
Acid	R – COOH	n — π*	205	1.5
Ester	R – COOR'	n — π*	205	1.5
Amide	R – CONH₂	n — π*	210	1.5

Uses in biology

UV/vis spectrophotometry can usefully be applied to proteins containing aromatic amino acids (see Topics L2 and M1) or porphyrins and to both DNA and RNA oligonucleotides (see Topics L2, M2, and *Table 3*), due to the UV absorption of the ring component of these molecules.

The most common use in biology is to measure sample concentrations. However, UV/vis spectrophotometry may also be used to monitor the effect of the environment on properties of biomolecules.

For example, the denaturation of double-stranded DNA and RNA (see Topics H1, H2, K2 and M2) may be followed by UV/vis spectrophotometry. Double-stranded nuclei acids are stabilized by hydrogen bonding and base-stacking interactions (see Topics H1 and H2). When heat is applied to a solution of a nucleic

Table 3. Wavelength maxima and extinction coefficients for common biological molecules

Molecule	λ_{max} (nm)	ε_{max} ($\times 10^{-3}$ cm^2.mol^{-1})
Tryptophan	280	5.6
	219	27.0
Tyrosine	274	1.4
	222	8.0
	193	48.0
Phenylalanine	257	0.2
	206	9.3
	188	60.0
Histidine	211	5.9
Adenine	260.5	13.4
NADH	340	6.23
	259	14.4
NAD$^+$	260	18.0
Guanine	275	8.1
Uracil	259.5	8.2
Thymine	264.5	7.9
Cytosine	267	6.1

acid, sufficient energy is supplied to disrupt these interactions, and the double helix is said to **melt**. If the absorbance of the sample is measured at 260 nm it is found to increase as temperature increases. This change in hypochromicity is due to chromophores becoming separated, as base-pairing interactions are broken. This assay can therefore be used to determine the stability of the system. Indeed it is common to quote the **melting temperature** (T$_m$) for a sequence when talking about helix stability. The melting temperature is the center of the sigmoidal change in absorbance with temperature (see *Fig. 5*), and is defined as the temperature at which half of the molecules present have melted. This information can also be used to determine the ratio of A : T (or A : U) and G : C base pairing in DNA (or RNA).

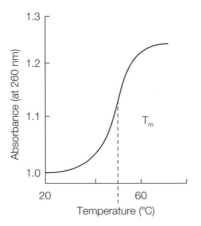

Fig. 5. The change in UV absorbance as a function of temperature for a short nucleic acid sequence.

Q3 FLUORESCENCE

Key Notes

Definition of fluorescence	Upon the absorption of light, electrons are promoted from the ground state to the excited state of a molecule. The excited state is short lived and the electrons gradually relax back to the ground state. The lower energy state may be regained by loss of energy as heat to the surrounding medium. Alternatively reradiation can occur with spontaneous emission of light. This light is called fluorescence.
Fluorophores	A fluorophore is the name given to a molecule, or part thereof, that gives rise to the fluorescence response. In nature the main fluorophores are the aromatic amino acids, flavins, vitamin A, and NADH. It is these groups that are monitored in fluorescence experiments on biological systems.
Quantum yield	Fluorescence is not the only process by which an excited state may be depopulated; other processes compete. These additional processes therefore cause a more rapid decay of the excited state than would be predicted if fluorescence alone was involved. The fraction of excited molecules that become deexcited by fluorescence is called the quantum yield and is symbolized, Φ_f.
Fluorescence quenching	The quantum yield of a fluorophore can be diminished or quenched by a range of processes. Fluorescence quenching can occur due to the close proximity of other fluorophores or due to the presence of paramagnetic species or heavy metal ions.
Uses in biology	Fluorescence may be used to determine sample concentration and to follow the course of a reaction. In addition the timescale of fluorescence is sufficiently fast that molecular rearrangements or molecular dynamics may be monitored.
Related topics	The electromagnetic spectrum (Q1) Ultraviolet-visible spectrophotometry (Q2).

Definition of fluorescence

Generally, at room temperature molecules are in their ground electronic state and vibrational level (see Topics Q1 and Q2). Absorption of energy at an appropriate wavelength results in electronic excitation to the **singlet state** and the upper vibrational level associated with this electronic transition. To regain the equilibrium or ground state, several processes may take place, enabling energy loss. The first event is that molecules in vibrational levels above the lowest in the excited state lose their excess energy (usually in the form of heat), moving to a lower vibrational level, but still in the excited electronic state. From this state spontaneous emission can occur, and the light that is emitted is **fluorescence** (see *Fig. 1*).

The outcome of a fluorescence experiment may be depicted as an **excitation spectrum** or an **emission spectrum**. In the former the fluorescence intensity

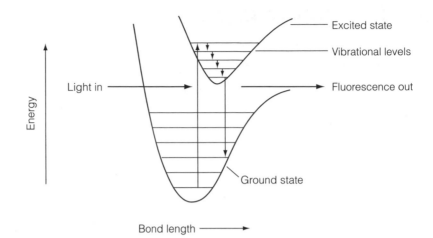

Fig. 1.   Processes which lead to fluorescence.

depends on the wavelength of the **exciting** light, in the latter it depends on the wavelength of the **emitted** light.

Fluorophores

A molecule, or part thereof, which gives rise to fluorescence is referred to as a **fluorophore**, in an analogous manner to chromophore in UV/vis spectrophotometry (see Topic Q2). In general, molecules containing conjugated pi systems will give rise to fluorescence. In nature these include molecules such as the aromatic amino acids, NADH, the flavins and vitamin A, for example. In addition, a large number of synthetic fluorophores have been developed, referred to as fluorescent probes, which are utilized in studies of binding to biological molecules which otherwise would not be readily analysed by fluorescence. See *Table 1* for examples of fluorescent probes and for characteristics of these and natural fluorophores.

Quantum yield

Fluorescence is not the only process by which the ground state for a molecule may be regained. Rather, there are several competing processes, all of which bring about a more rapid decay of the excited state than would be the case if fluorescence was the only process operating (see *Fig. 2*). If you consider the overall rate of the relaxation process, clearly this must depend on the rates of each of the individual processes. If we assume that all the events follow first order kinetics (see Topic P1) then we can express the rate constants as follows;

$$k = k_f + \Sigma k_i$$

Table 1.   Natural fluorophores and synthetic fluorescent probes

Fluorophore (natural and synthetic)	Emission properties	
	λ_{max} (nm)	Φ_f
Tryptophan	348	0.20
Tyrosine	303	0.10
Phenylalanine	282	0.04
Dansyl chloride	510	0.10
Ethydium bromide	600	~1

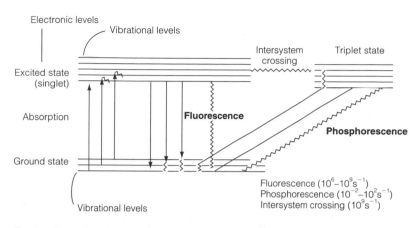

Fig. 2.   Examples of relaxation pathways and their timescales. (Reproduced from Biological Spectroscopy, I.D. Campbell and R.A. Dwek, Addison Wesley Longman 1984 with the permission of Professor R.A. Dwek.)

where k is the overall rate constant, k_i is the rate constant for each of the competing radiationless processes, and k_f the constant for fluorescence. The relaxation time, τ is defined as $1/k$.

The quantum yield, which is the fraction of molecules that have relaxed by fluorescence alone, is given by:

$$\Phi_f = \frac{k_f}{k_f + \Sigma k_i}$$

In terms of relaxation times the quantum yield may also be written as:

$$\Phi_f = \frac{\tau}{\tau_f} \qquad \tau_f = \text{fluorescence lifetime}$$

The quantum yield is a difficult parameter to measure and in practice is obtained by comparison to a standard sample.

Fluorescence quenching

As we have seen fluorescence is not the only relaxation process in operation for electronically excited molecules. The presence of other processes means a reduced fluorescence intensity, that is, the fluorescence is quenched. In some instances specific quenching process can be analyzed leading to useful structural information.

There are two types of quenching most routinely studied in biology; resonance energy transfer, and collisional energy transfer. Resonance energy transfer is a non-radiative process which occurs over relatively long distances (up to around 5 nm) between fluorophores. A detailed analysis of this process enables the distances between fluorophores to be measured.

Collisonal energy transfer occurs if a fluorophore collides with a paramagnetic agent, for example O_2 (see Topic F3), or a heavy ion, for example I^- or Cs^+, and then energy transfer, and thus relaxation, can be achieved by a process referred to as **intersystem crossing**, (see *Fig. 2*). In this instance the paramagnetic agent or ion is the quencher and analysis of this type of process can provide information regarding molecular dynamics.

Uses in biology

Fluorescence, just like ultraviolet absorption, may be used to monitor sample concentrations and follow reactions that result in a change in the nature or environment of the fluorophore. For example, the binding of nicotinamide adenine dinucleotide in both reduced and oxidized forms (NADH and NAD^+ respectively) to glutamate dehydrogenase, in the presence of a substrate analog, glutarate, may be followed by fluorescence studies. Such investigations have suggested that NAD^+ and glutarate bind to three of the six subunits of the dehydrogenase and bring about a conformational change which has consequences for the fluorescence quantum yield of NADH. Perhaps the most important application of fluorescence is to the study of molecular motions and molecular structure.

Fluorescence quenching may occur by the transfer of energy between fluorophores separated by a relatively large distance. A donor (excited system) and acceptor (in the ground state) is required, also the energy separations between the excited and ground state for each must be equivalent. The rate of energy transfer between donor and acceptor has been found to depend on $1/r^6$, where r is the distance between the fluorophores. It is therefore possible to, for example, monitor the distance between a fluorescence probe bound to a protein and amino acids in the active or binding site of that protein.

Q4 INFRARED SPECTROPHOTOMETRY

Key Notes

Molecular vibrations
At room temperature, molecules are in a constant state of vibration. The energy required to stretch or bend a bond is of the order of 10^3 to 10^5 Jmol^{-1}, which corresponds to the infrared (IR) region of the electromagnetic spectrum. Hence, infrared spectrophotometry is used mainly to differentiate between bond types, or more specifically between functional groups.

The infrared spectrum
The output of an infrared (IR) experiment is a plot of percentage absorbance or transmittance (of infrared radiation) against frequency. More usually wavenumbers are quoted rather than frequency. The wavenumber (cm^{-1}) is the reciprocal wavelength (cm). The absorbance or transmittance is not normally quantified, bands are simply labeled strong, medium, or weak.

Functional group banding
The bending and stretching vibrations for specific functional groups fall into narrow well defined regions and thus provide a diagnostic of compound class. In the case of organic/biochemical compounds these key bands fall in the region 1600–4000 wavenumbers (cm^{-1}). The region below 1600 cm^{-1} being referred to as the fingerprint region.

Uses in biology
The main uses of IR in biology are to study ligand binding, to probe hydrogen bonding interactions and, in special circumstances, molecular conformations.

Related topic
The electromagnetic spectrum (Q1)

Molecular vibrations

Under normal conditions of temperature and pressure, molecules are in a constant state of vibration. In the simplest case of a diatomic molecule, the atoms attached to either end of a bond may be likened to balls on the end of a spring, and Hooke's law may be applied to determine the energy (actually the frequency but this is proportional to the energy) required to stretch the bond (spring) (*Fig. 1*).

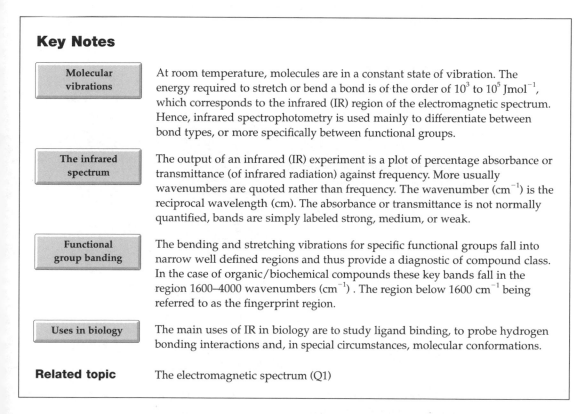

$$\upsilon = \frac{1}{2\pi}\sqrt{\frac{k}{\left(\dfrac{m_1 m_2}{m_1 + m_2}\right)}}$$

m_1, m_2 = masses of atoms on bond
k = force constant
υ = frequency

$\dfrac{m_1 m_2}{m_1 + m_2}$ = reduced mass = μ

Fig. 1. Hooke's law may be applied to atoms and bonds.

As can be seen the frequency of a stretching vibration is proportional to the square root of the bond strength, or force constant, divided by the reduced mass. This proportionality enables prediction of the relative positions of bands for particular bond types provided the strength of the bond is known and the mass of the attached atoms is known.

A simple diatomic molecule has only one mode of vibration, bond stretching. In a polyatomic system there are many vibrational modes, (3N – 6) to be precise, where N is the number of atoms. An example of the various modes of vibration is shown in *Fig. 2*.

Stretching vibrations

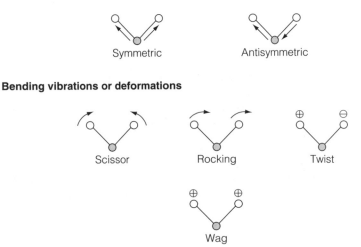

Bending vibrations or deformations

Fig. 2. *Some vibrational modes for a three atoms system.* ——, *bond;* ⊙○, *atoms,* +, *out of plane of paper;* –, *behind plane of paper.*

More energy is required to stretch a bond than to bend a bond and a symmetric stretch requires more energy than an antisymmetric stretch.

An important point to note is, for a vibration to be **detected** in the infrared spectrum, it must occur with a change in **dipole moment**.

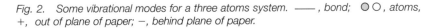

symmetric stretch – NO dipole moment change
NO IR response

We can readily use Hooke's law to predict;

$$\left.\begin{array}{l} \upsilon\, C = C > \upsilon\, C \!\!-\!\! C \\ \upsilon\, C = O > \upsilon\, C \!\!-\!\! O \end{array}\right\} \text{due to } k$$

$$\left.\begin{array}{l} \upsilon\, C \!-\! H > \upsilon\, C \!\!-\!\! C \\ \upsilon\, O \!-\! H > \upsilon\, C \!\!-\!\! O \end{array}\right\} \text{due to } \mu$$

The infrared spectrum

The result of passing an infrared beam, whose wavelength is systematically varied, through a sample is depicted as a plot of percentage transmittance or absorbance of light against wavelength or frequency. A change in absorbance or transmittance occurs when the frequency of a bond vibration matches that of the applied infrared beam. It is this superposition that is detected. An example of an infrared spectrum for the hydrocarbon Nujol, is presented in *Fig. 3*.

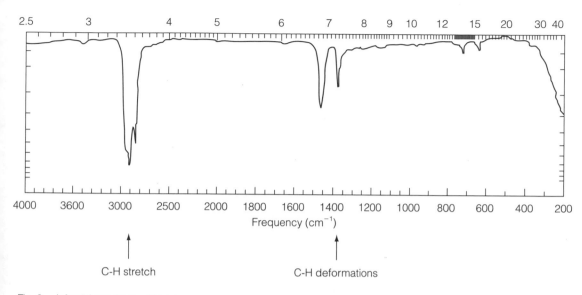

Fig. 3. Infrared spectrum of Nujol.

IR experiments may be performed in the gas, liquid, or solid phase. Most routinely solid samples are presented as suspensions in Nujol, or compressed in potassium bromide disks. Liquids or Nujol suspensions are presented to the IR beam by sandwiching them between disks (usually 2 cm diameter) made of sodium chloride crystals. Sodium chloride does not absorb in the infrared region of general interest. Although IR data can be quoted in frequency or wavelength terms, it is more common to quote wavenumbers;

$$\bar{\upsilon}$$

Wavenumbers being related to frequency as follows;

$$\bar{\upsilon} = \frac{\upsilon}{c}$$

c = speed of light

The absorbance or transmittance is only referred to as being strong (s), medium (m), weak (w), or variable (v).

Functional group banding

The generally accessible region of the infrared spectrum spans 4000–625 cm^{-1}. Routinely, however, it is only the region between 4000–1600 cm^{-1} that is interpreted as this is where the diagnostic bands for the main functional groups appear. The region below 1600 cm^{-1} is referred to as the fingerprint region and is dependent on the full structure and bonding details of the molecule and is potentially very complex.

Fig. 4 is a schematic representation of the spread of functional group absorptions. Clearly it is straightforward to differentiate between an alcohol and a ketone, or an amine and a nitrile. Upon close inspection it is also reasonably straightforward to differentiate between, for example, ketones and esters, and their carbonyl stretching vibrations are at slightly different frequencies.

Uses in biology

Uses for infrared spectrophotometry in biology have been significantly increased in recent years by the introduction of Fourier transform infrared (FTIR), an

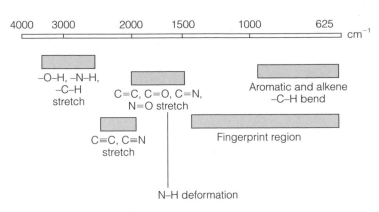

Fig. 4.   Schematic representation of positions of functional group absorption bands.

improved way of performing the basic IR experiment. The details of this technique are beyond the scope of this text.

A process which is readily monitored by IR is hydrogen(H)-deuterium (D) exchange and this is of relevance for biological systems such as proteins and nucleic acids, each of which have labile (exchangable) hydrogens.

The IR spectrum of a protein in D_2O (heavy water) solution is shown in *Fig. 5*. When the protein is fully folded, only surface residues are susceptible to H-D exchange. If we consider hydrogens attached to nitrogen, and note that an N-D bond is slightly stronger than an N-H bond, then we would expect a shift in the IR absorptions for those amine groups exposed to the solvent. The shift is not great but becomes more noticable as the protein unfolds (or denatures) and more amine groups become solvent accessible. IR is therefore useful to monitor denaturation–renaturation processes in proteins.

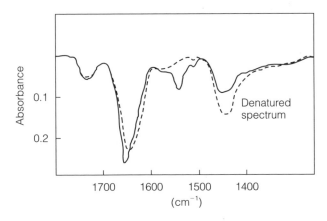

Fig. 5.   A typical infrared spectrum for a protein in D_2O solution fully folded (solid line) and denatured (dashed line). (Reproduced from Biological Spectroscopy, I.D. Campbell and R.A. Dwek, Addison Wesley Longman, 1984, with the permission of Professor R.A. Dwek.)

Q5 CIRCULAR DICHROISM

Key Notes

Circularly-polarized light

If two beams, or waves, of plane polarized light travel perpendicular to each other, with the same amplitude but differing in phase by 90 degrees, these waves superimpose to form a new wave which is said to be circularly polarized. If the amplitudes of the two waves differ, the resultant wave is elliptically polarized.

Definition of circular dichroism

If a molecule is optically active it interacts differently with left (**L**)- and right (**R**) circularly polarized light. If the absorption of light, by the molecule, differs between the two directions then the emerging light is elliptically polarized and circular dichroism (CD) is observed. Should the velocity of the light change then optical rotatory dispersion (ORD) is observed, which is related to refractive index.

Molar ellipticity

The result of a circular dichroism experiment is a measure of the difference in absorption (ΔA) of two perpendicular, circularly polarized beams of light. ΔA is related to the difference in extinction coefficients, via the Beer-Lambert law. The difference in extinction coefficients ($\Delta\varepsilon$) multiplied by 3300 is referred to as the molar ellipticity ($[\theta]_{\lambda}$) and this is often reported.

Chromophores for CD

The optical activity required for CD experiments arises in functional groups which possess an asymmetry with respect to the charge transition which occurs on moving electrons from the ground to excited state. Such groups are referred to as chromophores. The sign and the magnitude of the optical activity can be affected by the local environment. The amide bond and the disulphide bond of cystines are intrinsically optically active. The individual bases of nucleic acids are not optically active but optical activity is induced when these are found in oligonucleotides.

Uses in biology

Circular dichroism may be used to monitor ionization and ligand binding processes in proteins, to investigate conformation changes in nucleic acids and to establish the presence of secondary structure elements in peptides and proteins.

Related topics

Optical activity and resolution (E3)
Hydrogen bonds (H1)
Hydrophobic interactions (H2)

The electromagnetic spectrum (Q1)
Ultraviolet-visible spectrophotometry (Q2)

Circularly-polarized light

The electric component of light (see Topic Q1) may be described as a wave that is propagated in the y-direction and oscillates, sinusoidally in the yz plane (see *Fig. 1*). It is equally correct to depict this oscillation taking place in the yx plane, or indeed any direction perpendicular to the propagation axis. This is what occurs in unpolarized light. If all but one of these oscillations were to be blocked (by the use of a polarizer) then light would be produced with a single oscillation path, this is **plane-polarized light** (see Topic E3). **Circularly-polarized light** is produced when two beams of plane-polarized light are generated which oscillate perpendicular to each other and 90 degrees out of phase with each other (see *Fig. 1*). These waves superimpose to produce a 'wave' which is helical/circular in nature. If however the two beams are unequal in amplitude the superposition results in a wave which has an elliptical shape; hence **elliptically polarized light**.

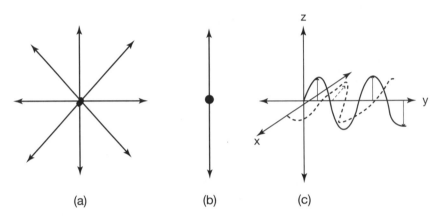

Fig. 1.   Representations of the path of the electric component of light. (a) Unpolarized light. (b) Light polarized in the z axis. (c) The wave motion of z polarized light in the zy plane (the dashed line indicates y propagated light in the yx plane)

Definition of circular dichroism

The sum of two circularly polarized beams of light rotating in opposite directions (**L** and **R**) is a beam of plane polarized light; the descriptions are interchangeable (see *Fig. 2*). This concept is important in understanding how **circular dichroism** (CD) comes about. A compound that is optically active will interact differently with light circularly polarized in the **L**-direction than with light in the **R**-direction. This differing interaction may manifest itself in a variation in the absorption of the two beams by the sample; i.e. $A_L \neq A_R$, this is **dichroism**. As absorption is related to extinction coefficient ε (see Topic Q2) this means that $\varepsilon_L \neq \varepsilon_R$. Alternatively the velocity of the beams passing through the sample may differ; i.e. the refractive index (n) is different in the two directions. This is the basis of **optical rotatory dispersion** (ORD).

If $A_L \neq A_R$ (or $\varepsilon_L \neq \varepsilon_R$) then the resultant, or combination, of the **L** and **R** beams will no longer be a beam of plane polarized light but rather **elliptically polarized light** (see *Fig. 2*) and circular dichroism is observed. If $n_L \neq n_R$ then the sum of **L** and **R** will still be plane polarized but will be rotated away (by α degrees) from the direction the light would travel in the absence of the sample (see *Fig. 2*); this is the basis of polarimetry (see Topic E3).

Molar ellipticity

In performing a CD experiment the absorption difference, ΔA, where ΔA = $A_L - A_R$, is the parameter observed. However, CD spectra are often plotted with the extinction coefficient difference, (Δε), in the y axis, or the related quantity,

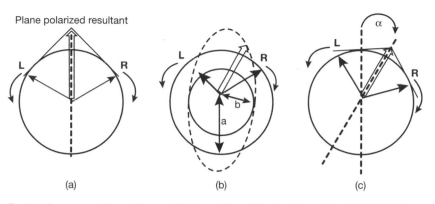

Fig. 2.   A representation of the combination of **L** and **R** circularly polarised light (a) in the absence of an optically active sample, and in the presence of an optically active sample when (b) circular dichroism is the result, and (c) when optical rotatory dispersion is the result.

the **molar ellipticity** ($[\theta]_m$ or $[\theta]_\lambda$), where θ is measured in degrees and λ is the wavelength of light used to make the observation. $[\theta]_\lambda$ and $\Delta\varepsilon$ are related as follows:

$$[\theta]_\lambda = 3300\ \Delta\varepsilon$$

The molar ellipticity is defined as:

$$[\theta]_\lambda = (\theta_{obs} \cdot MW)/100\ Lc$$

where θ_{obs} is the observed ellipticity in degrees (see *Fig. 2*), L is the path length in decimeters, and c is the concentration in grams per milliliter. The ellipticity may be determined from measurements 'a' and 'b' using *Fig. 2b* and the following relation;

$$\tan^{-1}\theta = b/a$$

In practice rather than using two separate circularly polarized light sources it is possible to use a crystal that can be 'tuned' to pass either **L** or **R** light by the changing of an applied voltage; this is called a **Pockels cell**. The wavelength of the applied light can be readily changed (generally in the region 200–700 nm) and ΔA is detected as a function of wavelength.

Typically spectra are broad and may display positive and negative components (see *Fig. 3*).

Chromophores for CD

A molecule is optically active (see Topic E3) if it can undergo an electronic transition in which the charge displacement has both a finite electric (from a linear displacement of charge), and finite magnetic (from the rotating electric current), dipole moment. A charge displacement of this type follows a helical path. An optically active molecule will therefore interact differently with **L** and **R** circularly (helically) polarized light. Such molecules, or parts thereof, are referred to as **chromophores** (see Topic Q2) and are generally asymmetric (see Topics E2 and E3). In biological systems natural chromophores include the amide bond (see Topics J4 and K1) and the disulphide link of cystines; these may be monitored in the range below 240 nm, and 240–360 nm, respectively. Neither the aromatic side chains of amino acids nor the heterocyclic (aromatic) bases of nucleic acids are intrinsically optically active. Optical activity may be induced in these residues by

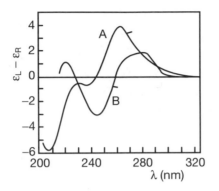

Fig. 3. Sketch of a CD spectrum produced from B-type and A-type DNA.

other environmental factors. For example, nucleic acids display optical activity when they are oligomerized as in strands of DNA or RNA (see Topics H2, K2, and M2). In these long chain molecules the heterocyclic bases stack and this brings about a splitting in the $n - \pi^*$ transition and this manifests itself as a positive and negative band in the CD spectrum (in the region 200–320 nm) (see *Fig. 3*). In general if two chromophores are close in space the transition dipole moments interact and this results in two CD bands of opposite sign (or sense).

Uses in biology   Circular dichroism may be used to monitor the ionization of, for example, car-boxyl groups in the active sites of enzymes. Measurement of $[\theta]_\lambda$ as a function of pH for an enzyme in the presence and absence of a substrate analog or inhibitor will produce two different curves if the ionization capability of one of the carboxyl groups is affected by ligand binding.

The conformation of DNA or RNA molecules (see Topic K2) may be established using CD spectrophotometry. As already mentioned nucleic acid bases are intrinsically optically inactive, but they are induced to have optical activity when in oligonucleotides, however the conformation of the oligomer determines the actual CD response. DNA is generally found in the form of a double helix with two oligonucleotide strands wound around each other to form a right-handed helix classified as B-type (see Topic K2). In this conformation the CD spectrum comprises of a negative band at around 240 nm and a positive band at around 270 nm. The shape and band maxima and minima change as the conformation is changed to A-type (see Topic K2). This is a direct result of changes in the base-stacking arrangement between the two conformations (see *Fig. 3*). Similarly the presence of regular secondary structure elements in peptides and proteins (K1) produce characteristic CD profiles (see *Fig. 4*).

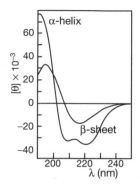

Fig. 4.   Sketch of the CD spectra of protein with significant α-helical content, or β-strand content.

Q6 NUCLEAR MAGNETIC RESONANCE SPECTROSCOPY

Key Notes

Nuclear spin	In nuclear magnetic resonance (NMR) spectroscopy the property which is monitored is that of nuclear spin. Those nuclei that have spin also posses a magnetic moment. It is the effect on this magnetic moment of an external magnetic field which is monitored in NMR spectroscopy. In a magnetic field the magnetic moment can adopt two orientations which differ in energy. The energy required or emitted in switching from one orientation to another is in the radio frequency region of the electromagnetic spectrum.
Nuclei with spin	Many atoms have spin and are therefore detectable in an NMR experiment. Amongst the most important is the most abundant isotope of hydrogen, ^{1}H; in this context referred to as a proton. Other useful nuclei include ^{31}P and ^{19}F. The most abundant isotope of carbon, ^{12}C, does not have a spin and therefore is NMR silent. However, ^{13}C does have a spin and this nucleus is utilized in modern NMR studies.
The NMR spectrum	Modern NMR spectrometers are referred to as Fourier Transform (FT) instruments. The signal detected at the end of an experiment is a function of time. This signal is Fourier transformed to produce a signal which is a function of frequency and is the NMR spectrum. Specific features which are extracted from the spectrum are the intensity of peaks; their integrals, their positions; their chemical shifts, and fine structure on peaks; multiplicities and coupling constants.
Integrals	One of the key features of NMR is that the area under each peak in a spectrum is proportional to the number of nuclei absorbing or emitting at that frequency. The area under each peak is referred to as the integral and is used to determine the numbers of nuclei at chemically different sites in a molecule.
Chemical shifts	Peaks appear in the NMR spectrum, corresponding to the absorption of different frequencies of radiation. The difference in frequencies in, for example, protons within a molecule is very small, but is dependent on the external magnetic field. To overcome these problems, a dimensionless property, the chemical shift, is quoted. This is the difference in frequencies between sample protons, and those of some standard, divided by the carrier (machine operating) frequency. The chemical shift is quoted in parts per million (p.p.m.).
Coupling constants	In many instances, NMR peaks have fine structure. This is due to the interaction between nuclear spins (which are nuclear magnets) within the molecule. The number of lines, or multiplicity, can be interpreted to determine the number of interacting spins within the molecule. The size of the line separation within the multiplet, the coupling constant, relates to the energy of

interaction which can in turn be related to the number of intervening bonds and angles between these.

Relaxation

NMR excited states are particularly long-lived, consequently it is possible to monitor the processes *via* which the ground state is regained, or relaxation is said to occur. In utilizing or monitoring relaxation events, information regarding molecular flexibility and internuclear distances may be determined.

Applications in biology

NMR may be used to investigate the pH dependence of biomolecular processes and the pK_a's of specific residues. It can be used to monitor the kinetics of ligand binding. Under ideal conditions (high biomolecule concentration and pH and temperature stability) NMR can also be used to determine macromolecular structure.

Related topic

The electromagnetic spectrum (Q1)

Nuclear spin

If a nucleus has spin, then it has an intrinsic angular momentum. This coupled with the fact that every nucleus is positively charged, means that nuclei with spin have a magnetic moment. This magnetic moment can adopt different orientations, an energy difference arises between these orientations when the nucleus is in a magnetic field.

Nuclei with spin can take up $2I + 1$, where I is the spin, possible orientations relative to the external magnetic field. Therefore, when $I = \frac{1}{2}$, as for ^1H and ^{13}C, two orientations are possible. The lower energy orientation is labeled α, the higher β. It is transitions between these levels that are monitored (see *Fig. 1*). The size of the energy gap is dependent, amongst other things, on the strength of the magnetic field. The currently available magnets range from 1.4 to 17.5 tesla. The energy required therefore for a transition to occur is in the radio frequency region of the EM spectrum (see Topic Q1).

As the energy difference is very small, under normal conditions of temperature and pressure, there is a small spin population difference between the energy levels, with a slight excess in the α, spin state. It is the population difference that we consider as being manipulated in the NMR experiment, as a result NMR is a relatively insensitive technique.

Nuclei with spin

Depending on their atomic and mass numbers (see Topic A1) nuclei may have half integral, integral or zero spin. For example, ^1H, ^{19}F, ^{31}P and ^{14}N, have spin ($\frac{1}{2}$, $\frac{1}{2}$,

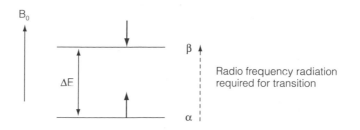

Fig. 1. Spin ½ nuclei can adopt two orientations. In a magnetic field, B_0, the lower energy orientation is labeled α, or $+\frac{1}{2}$, the higher β or $-\frac{1}{2}$. The energy required to flip between these spin states is in the radio frequency region of the electromagnetic spectrum.

½ and 1, respectively), however, ^{12}C and ^{16}O, the most abundant isotopes of key elements in organic chemistry and biology, do not have spin. Other isotopes which may be routinely observed by NMR include ^{13}C and ^{15}N (both spin ½), although their low relative natural abundancies make data acquisition more time consuming. Of the nonradioactive isotopes ^{1}H is the most sensitive nucleus for NMR study.

The NMR spectrum

Most of the spectrometers in use today are so-called **Fourier transform** (FT) instruments. This means that all of the frequencies required to excite the different nuclei (these may all be ^{1}H) are supplied at once in a short high energy burst, referred to as a **pulse**. The loss of this energy to regain the ground state is monitored as radio signals by a set of coils surrounding the sample vessel. The signal thus detected is a function of time and is referred to as the **free induction decay** (fid) (see *Fig. 2*). This process may be repeated numerous times with the results being coadded in the spectrometers computer memory. The traditional spectrum is produced by Fourier transforming this signal to produce a frequency-dependent signal (see *Fig. 2*).

The spectrum is then plotted on paper and a scale is drawn on automatically. In reporting the results of an NMR spectrum, several features of the spectrum are given; the area under each peak or the **integral**, the positions of each peak or the **chemical shift** (δ), and the fine structure of each peak, the **multiplicity** and **coupling constant** (J) (see *Fig. 3*).

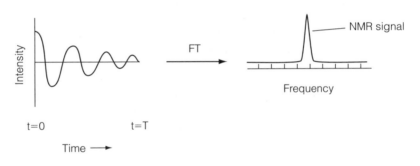

Fig. 2.   *The free induction decay (fid), a time dependent signal, is Fourier transformed into the frequency dependent spectrum.*

Integrals

One of the most useful features of NMR is that the area under each peak is directly proportional to the number of nuclei in that particular chemical environment. Consequently, in the ^{1}H NMR spectrum of ethanol, CH_3CH_2OH, three peaks would be detected for the three chemically different hydrogens, the area under each peak would integrate for a ratio of $3:2:1$. The integral is generally shown as a stepped line above each peak (see *Fig. 3)* or in numerical form near to the peak. This is clearly a useful piece of information in assigning NMR spectra.

Chemical shifts

Nuclei in different chemical environments will experience slightly different magnetic fields and their signals will therefore be at slightly different frequencies. To avoid working with small numbers and to overcome the fact that the resonance frequency is dependent on the external magnetic field, a dimensionless parameter is utilized, the **chemical shift** (δ) (*Fig. 4*).

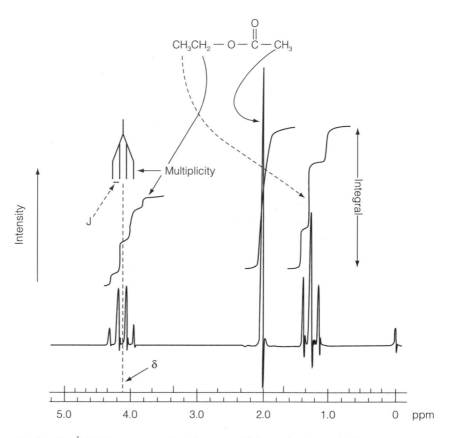

$$CH_3CH_2 - O - \overset{\overset{\displaystyle O}{\|}}{C} - CH_3$$

Fig. 3.   The ^1H NMR spectrum of ethylacetate with integrals, chemical shifts, and multiplicities indicated. δ, chemical shift (in ppm); J coupling constant (in Hz).

The size of a chemical shift is dependent on a number of factors, but all of these relate to the degree of shielding or exposure to the applied magnetic field afforded to the nucleus by 'surrounding' electrons. Electrons circulating around a nucleus set up a secondary magnetic field which can 'shield' the nucleus from the external magnetic field (B_0). The effective field at the nucleus (B_{eff}) being smaller than B_0, means that the nucleus has a smaller chemical shift than would have been predicted. As the nature and arrangement of bonds near a particular nucleus can effect the electron density about that nucleus, so also can these effect the chemical shift. Therefore, in assigning chemical shifts we are establishing the types and locations of the various nuclei.

$$\delta = \frac{\upsilon_{sample} - \upsilon_{reference}}{\text{carrier frequency}} \times 10^6$$

δ   = chemical shift in parts per million (ppm)
υ_{sample} = resonance frequency of nucleus in sample (Hz)
$\upsilon_{reference}$ = resonance frequency of nucleus in a reference compound (Hz)
carrier frequency = operating frequency of spectrometer for particular isotope (MHz)

Fig. 4.   Definition of the chemical shift, δ.

The key influences on chemical shifts are the electronegativity (inductive effect, see Topic I3) of atoms attached to the observed nucleus, and the proximity of delocalized, or delocalizable, pi or lone pair electrons (see Topics J3 and L1). Of course the inductive effect decreases with number of intervening bonds and this is observed in the NMR response. Therefore, in the ^{1}H NMR spectrum of ethanol one would predict, on the basis of the electron withdrawing power of oxygen, that the chemical shifts would increase in the order $\underline{CH_3} < \underline{CH_2} < O\underline{H}$, and this is what is observed.

When pi (π), or pi (π) and lone pair electrons are present in a molecule this can cause an additional magnetic field to form (a **molecular circulation of electrons**), which may oppose or align with the applied magnetic field (see *Fig. 5*). There is an **anisotropy** about this additional magnetic field; that is, the effect is not the same in all directions. Such magnetic anisotropies occur in all multiply-bonded systems and add to the shift induced by attached electron withdrawing groups.

There are many other factors which contribute to the chemical shift, but to a lesser extent. A full interpretation of these can aid structure analysis.

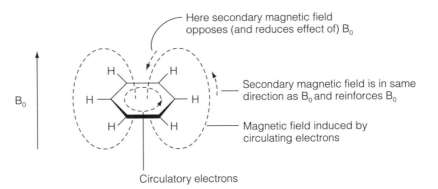

Fig. 5. Schematic representation of the magnetic field which surrounds benzene. Note in certain regions this field is aligned, and therefore enhancing the applied magnetic field, in other regions it is opposing and therefore reducing the magnetic field experienced at the nucleus. B_0, external magnetic field.

Coupling constants

Nuclei with spin separated by up to four (in some instances five) bonds interact with each other via their bonding electrons; in the same way that a row of magnets would exert a force on each other. The result of such interactions is the splitting of lines (fine structure) at the chemical shifts of the interacting or **coupled nuclei**. The number of lines resulting, the **multiplicity**, and the separation between these, the **coupling constant (J)**, may be interpreted to aid peak assignment and to determine molecular geometry.

A simple illustration of how splitting occurs may be provided by the consideration of a system of two spin ½ nuclei, A and X. If these nuclei are far apart in terms of bonds, then they will not couple, and the NMR spectrum will simply consist of two lines; one at the chemical shift of A and one at the chemical shift of X. If however these nuclei are close through bonds, they will each see the other's α and β spin states (see *Fig. 6*). This results in two transitions of differing energies (therefore differing frequencies) being possible for each nucleus. Consequently, centered at each of the chemical shifts will be two lines, of equal intensity (as each transition is equally probable) separated by the energy difference, in frequency

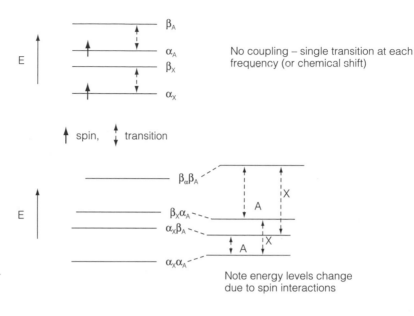

Assuming A has a larger chemical shift than X

No coupling – single transition at each frequency (or chemical shift)

↑ spin, ↕ transition

Note energy levels change due to spin interactions

$\alpha = \text{spin} +\frac{1}{2}$, $\beta = \text{spin} -\frac{1}{2}$

Fig. 6. Schematic representation of the energy level separations, and nuclear spin transitions, for two spin ½ nuclei, A and X.

terms, between the transitions. This energy, or frequency difference is the coupling constant (J).

Obviously the picture becomes somewhat more complicated when more nuclear spins are involved, however for the simplest cases, and with spin ½ nuclei, the **n + 1** rule and **Pascal's triangle** may be used to predict the number of lines and the relative intensities of lines in a multiplet. The n + 1 rule simply states that when a nucleus is coupled to n other nuclei (which must all be equivalent to each other), then the peak for the observed nucleus will be split in to n + 1 lines, with intensities given by Pascal's triangle (see *Fig. 7*).

The size of the coupling constant (J) is dependent on the number of bonds between the coupling nuclei, decreasing as the number of bonds increases. It is also dependent on the angles between the bonds connecting the nuclei. In this context the most useful coupling to measure is that over three bonds; the so called **vicinal** coupling as this can be related to the **dihedral** or **torsional** angle between the bonds *via* the **Karplus** equation;

$$^3J = A\cos^2\theta + B\cos\theta + C$$
$$\theta = \text{dihedral angle; A, B, C} = \text{constants}$$

Relaxation

The process *via* which nuclear spins regain the ground state following the application of a radio frequency pulse is referred to as **relaxation**. There are actually a number of mechanisms by which nuclear spins may relax and these can operate at the same time, but to different extents. These mechanisms are reasonably well understood and hence it is known how they relate to **motions** within molecules and **internuclear separation**. Experiments have been devised which enable quantifica-

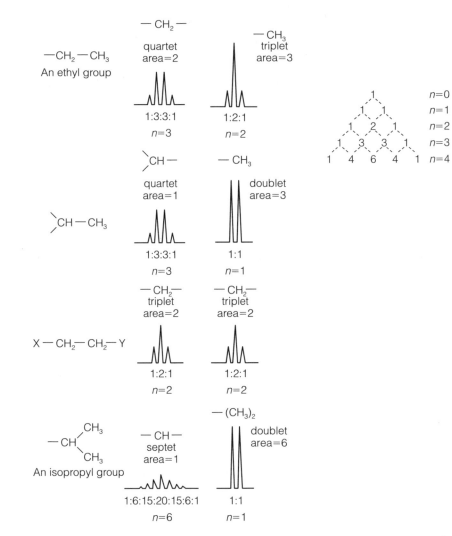

Fig. 7.   *Pascal's triangle and examples of common splitting patterns for a nucleus coupled to spin ½ nuclei. n, number of (equivalent) adjacent nuclei ('H's). Reprinted, with permission, from* Organic Structures from Spectra. Field *et al., copyright John Wiley and Sons Limited.*

tion of the contributions of the various processes. These experiments are particularly useful with large biological molecules, as for the most part to function properly they need to flex. The ability to determine internuclear separations, *via* measurement of the relaxation phenomenon called the **nuclear Overhauser effect (nOe)** enables the structure of these and smaller molecules to be determined in solution to compliment, or in the absence of, X-ray (solid state) data.

Applications in biology

NMR has numerous applications in biology. Indeed many of the recent advances in the technique have arisen as a result of their potential utility in biology. Many features of biomolecular systems and processes may be analyzed using this technique, with the proviso that the molecules under investigation are present at high concentrations (as NMR is quite insensitive) and, if possible, are stable over a range of pH and temperature.

It is possible to monitor the ionization of amino acid sidechains in proteins by ^{1}H NMR, to monitor the interaction between proteins and ligands (determining binding constants), and to determine the 3-dimensional shape of a biomolecule (with a molecular weight limit, depending of other features of the protein, of 50–100 kDa). To determine the structure of a large biomolecule in solution first requires that the proton spectrum be fully assigned. This is now relatively routinely achieved using advanced techniques such as 2–dimensional and 3–dimensional NMR. Each of these can benefit from the presence of both ^{13}C and ^{15}N at abundances well in excess of their natural levels. Advances in molecular biology techniques have made the production of isotopically enriched biomolecules feasible. Having assigned the proton chemical shifts measurement of **nOe** constraints permits interproton distances, hence 3-dimensional structures to be determined.

NMR studies of very large biomolecules can be fruitful when they involve, for example, the binding of small ligands. In such studies, features of the large molecule are determined indirectly, by monitoring the effects of binding on the NMR spectrum of the ligand.

Over and above structure-function analyses there are many NMR applications which utilize nuclei other than protons to monitor biological processes. For example, changes in intracellular pH may be monitored by following the ^{31}P spectrum of the inorganic phosphate present in the cells both *in vivo* (see *Fig.8(a)*) and *in vitro* (see *Fig. 8(b)*).

(a)

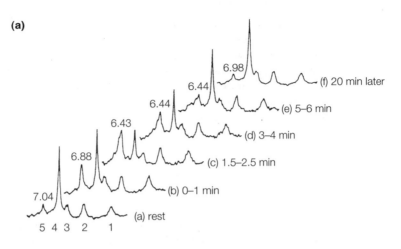

(b)

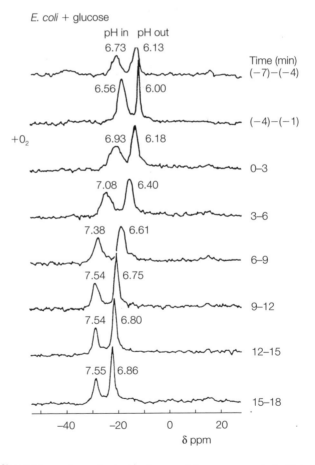

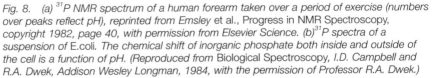

Fig. 8.   (a) ^{31}*P NMR spectrum of a human forearm taken over a period of exercise (numbers over peaks reflect pH), reprinted from Emsley* et al., *Progress in NMR Spectroscopy, copyright 1982, page 40, with permission from Elsevier Science. (b)* ^{31}*P spectra of a suspension of E.coli. The chemical shift of inorganic phosphate both inside and outside of the cell is a function of pH. (Reproduced from* Biological Spectroscopy, *I.D. Campbell and R.A. Dwek, Addison Wesley Longman, 1984, with the permission of Professor R.A. Dwek.)*

Q7 MASS SPECTROMETRY

Key Notes

Basic experiment

In mass spectrometry (MS) the species detected is an 'ion' formed by the bombardment of a sample with, in the simplest experiment, high energy electrons. As the resultant molecular ion is generally not stable, the ionized molecule can decompose to produce daughter ions. These breakdown products enable structural analysis.

Ion analysis

The simplest ion separation device is referred to as a magnetic sector device. However, for accurate molecular mass determination, the basic instrument set up is called a double focusing mass spectrometer, which utilizes an electrostatic and magnetic analyzer.

The mass spectrum

The output of a mass spectrometric analysis is generally displayed as a plot of relative abundance against the mass/charge (m/z) ratio. Alternatively, a listing may be presented of relative abundance and m/z ratio.

Ionization techniques

Ionization techniques fall into two categories; hard and soft. The most common technique for ion production is electron impact (EI), this is a hard ionization method. Other routine techniques include chemical ionization (CI) and fast atom bombardment (FAB). FAB-MS is generally used for polar or large molecules and is a soft ionization method. Many more ionization methods have been developed which are of particular use in biology.

Related topics   The periodic table (A1)   Isotopes (A3)

Basic experiment   As the name suggests mass spectrometry (MS) is primarily used to determine the molecular mass of compounds. Although it is not really a spectroscopic technique (as it does not involve the absorption or emission of electromagnetic radiation (see Topic Q1)) it is generally discussed alongside such techniques.

In MS the key event is the formation of an **ion** from (in general) a neutral molecule;

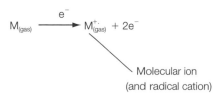

$$M_{(gas)} \xrightarrow{\ e^-\ } M^{+\cdot}_{(gas)} + 2e^-$$

Molecular ion
(and radical cation)

The first ion formed, in the simplest process, is referred to as the **molecular ion** as this ion has essentially the same molecular weight as the neutral molecule, the electron lost having negligible mass.

Under normal circumstances however, the molecular ion is usually very unstable and rapidly decomposes to produce **daughter** ions, **neutral molecules** and/or **free radicals**. The latter two species cannot be detected directly.

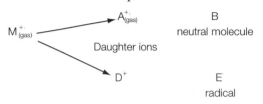

where A, B, D, and E simply represent species of lower molecular weight than the molecular ion.

Analysis of these fragmentation ions can aid the determination of not only molecular composition but also molecular structure.

The basic requirements for a mass spectrometer are therefore an ionization source, a method by which ions may be separated on the basis of their mass, and of course a detection procedure. The ionization source is perhaps the most important feature as this will determine the type of information that may be gained for a sample.

Ion analysis

The simplest instrument for ion separation is the **magnetic sector device** (see *Fig. 1*). This instrument relies on the fact that a charged species propelled through a magnetic field can be deflected by that field, in a manner that depends on the mass of the species and its charge, as described by the following expression;

$$\frac{m}{z} = \frac{B^2 r^2}{2V}$$

V = accelerating voltage; z = charge (generally +1); m = mass of ion; B = magnetic field strength; r = radius of the flight path of the ion.

Clearly by varying the magnetic field strength or the diameter of the flight path the device may be tuned to any molecular weight.

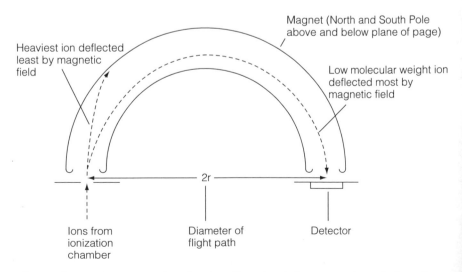

Fig. 1. Schematic representation of a magnetic sector device. - - -▶ *, ion path through magnetic field B.*

There is a drawback however in that this instrument only enables ions to be differentiated on the basis of integral mass units. The problem that this can cause is illustrated by considering a mixture of carbon monoxide, ethene, and nitrogen. Each of these molecules has an integral molecular mass of 28 units, therefore differentiation between these is not possible using a magnetic sector device. However, if we consider the accurate mass of each of the atoms, we find that carbon monoxide has a mass of 27.9949, ethene 28.0313, and nitrogen 28.0061 (see *Table 1*). Clearly a higher resolution device would enable unambiguous analysis. Such a device is referred to as a **double focusing** mass spectrometer (see *Fig. 2*)

The electrostatic analyzer attracts the ions deflecting their flight path, low molecular weight ions being deflected more than high molecular weight species. Therefore, molecular weight selection and detection may be fine tuned.

Table 1. Exact masses of some important isotopes

Element	Atomic weight	Isotope	Mass
Hydrogen	1.00797	^{1}H	1.00783
Carbon	12.01115	^{12}C	12.00000 (standard)
Nitrogen	14.0067	^{14}N	14.0031
Oxygen	15.9994	^{16}O	15.9949
Phosphorus	30.974	^{31}P	30.9738
Bromine	79.909	^{79}Br	78.9183
		^{81}Br	80.9163

The mass spectrum

The most common type of output from a mass spectrometric experiment is a spectrum. This is not like any of the spectra produced by other techniques in this section, as the x-axis is not a function of frequency or wavelength, but is simply the mass-to-charge ratio for the ion detected.

The abundance of a particular ion, which reflects its stability, is not presented in absolute terms, but relative to the most abundant ion; which is given an abundance of 100% (see *Fig. 3*). Often rather than, or in addition to, a spectrum a peak listing is presented.

The highest molecular weight ion is generally the molecular ion, if it survives long enough to be detected. Peaks can sometimes be detected at higher molecular weights if the sample becomes protonated, or it is ionized in a matrix, as in fast

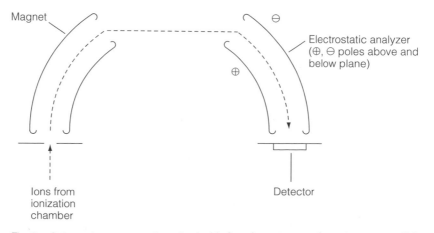

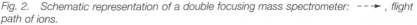

Fig. 2. Schematic representation of a double focusing mass spectrometer: – – ► *, flight path of ions.*

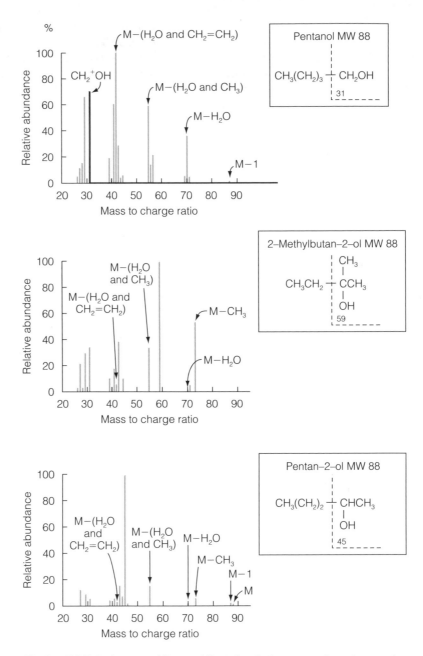

Fig. 3. EI-MS for isomers of Pentanol illustrating that mass spectrometry can also provide
structural data. Reprinted by permission of John Wiley and Sons Inc. from Spectrometric
Identification of organic compounds, 6th edn. Silverstein and Webster. Copyright © 1998,
this material is used by permission of John Wiley and Sons Inc.

atom bombardment mass spectrometry (FAB-MS). Of course isotopes will also be
observed in their natural abundance (see Topic A3).

Ionization   There are now a huge number of ionization techniques. The simplest, and the one
techniques   most routinely used by organic chemists, is the **electron impact (EI)** method. In

EI-MS a heated filament is the electron source, and these electrons are accelerated across a large potential difference and bombard sample molecules which are in the gas phase. In general, the electrons are moving so fast that they cannot be captured, but rather knock off an electron from the sample, with the production of a radical cation, as described above. This is the **molecular ion**, it usually gains so much energy from this process that bonds start to break and the molecule starts to decompose. In some instances it is possible to predict the fragmentation pattern, by considering the bond strength and the stability of the ion product, and therefore fragmentation patterns, this can enable the determination of functional group positions or branch points (see *Fig. 3*).

Chemical ionization (CI-MS) requires the use of a carrier gas. Methane or ammonia is bombarded with electrons and ionized. The ionization products may collide with each other to produce further ions, and it is these that are used to bombard sample molecules (see *Fig. 4*). This is a much less energetic process than EI and hence the molecular ion is somewhat more stable, more long-lived and therefore detected with a higher relative abundance than in EI-MS.

$$CH_4 \xrightarrow{e^-} CH_4^+ + 2e^- \longrightarrow CH_3^+ + H$$

$$\downarrow CH_4$$

$$CH_5^+ + CH_3$$

$$CH_5^+ + M_{(gas)} \longrightarrow MH^+ + CH_4$$

(Molecular weight +1)

Fig. 4.   The ionization of a carrier gas prior to sample bombardment.

Note that, dependent on the carrier gas is the charge of the ion detected; with methane positive ions are formed, with ammonia negative ions.

Fast atom bombardment (FAB-MS) involves the use of neutral atoms that have a high kinetic energy. Such atoms are formed by the bombardment of stationary atoms by rapidly moving ions of the same element. Upon collision momentum is transferred and a fast moving atom is produced. It is this species which is then in collision with sample molecules. In FAB-MS the sample is presented 'dissolved' in a matrix, generally of glycerol. The first ionized species usually has the molecular weight of the protonated molecule plus [92n], where 92 is the molecular mass of glycerol and 'n' is the number of molecules of glycerol associated with the protonated molecular ion;

$$\overrightarrow{Xe^+} + Xe \longrightarrow Xe + \overrightarrow{Xe^+}$$

Fast moving   Fast moving
Xenon ion   Xenon atom

as this is a soft ionization technique fragmentation is quite limited.

FAB-MS is particularly useful for studies of polar molecules or molecules of high molecular weight, hence it is particularly useful in studies of biological molecules.

Many other techniques have been devised which have applications in biochemistry or similar specialist areas. For example, electrospray ionization mass spectrometry (ESI-MS) is now being widely used. Again this is a soft ionization technique leading to molecular ions, and multiply charged variants, but little fragmentation.

FURTHER READING

This text is primarily aimed at first year undergraduates studying biology, biochemistry, or related subjects, with chemistry as a subsidiary course. Although the text is likely to be a useful companion for studies in latter years of such courses it is unlikely that readers of this book will require an indepth knowledge of many of the topics we have included, or require to consult original articles. Consequently in compiling a 'further reading' list we have only included textbooks. There are very many textbooks which cover, in a comprehensive manner, organic, inorganic, and physical chemistry as separate topics. For the most part these are not included here, the reader is simply advised that many excellent texts exist. Rather textbooks are identified which have useful introductory material only, or have been written specifically with biological sciences students in mind. The only exception to this is the in the speciality of spectroscopy. As spectroscopic techniques are likely to be important throughout biological sciences courses a number of books dealing with this subject matter are included. This is by no means an exhaustive list but does include many of the texts regularly referred to by the authors in their teaching of chemistry to biology and chemistry students.

General reading

Bettelheim, F.A. and March, J. (1995) *Introduction to General, Organic and Biochemistry*, 4th int. edn. Saunders College Publishing, Forth Worth, USA.

Buckberry, L. and Teesdale, P. (2001) *Essentials of Biological Chemistry*. Wiley, Chichester, UK.

Caret, R., Denniston, K. and Topping, J. (1997) *Principles and Applications of Inorganic, Organic and Biological Chemistry*, 2nd edn. McGraw-Hill, UK.

Gebelein, C. (1997) *Chemistry and Our World*. McGraw-Hill, UK.

Holum, J.R. (1994) *Fundamentals of General, Organic and Biological Chemistry*, 5th edn. John Wiley & Sons, Chichester, UK.

Sackheim, G.I. and Lehman, D.D. (1994) *Chemistry for the Health Sciences*. Macmillan, New York, USA.

Seager, S.L. and Slaubaugh, M.R. (1994) *Chemistry for Today – General Organic and Biochemistry*, 2nd edn. West Publishing, New York, USA.

Stoker, H.S. (2001) *General, Organic and Biological Chemistry* 2nd edn. Houghton Mifflin, Boston, USA.

More advanced reading

Sections B and D

Patrick, G.L. (1997) *Beginning Organic Chemistry 1*. Oxford University Press, Oxford, UK.

Patrick, G.L. (1997) *Beginning Organic Chemistry 2*. Oxford University Press, Oxford, UK.

Section G

Huheey, J.E. (1983) *Inorganic Chemistry – Principles of Structure and Reactivity*, 3rd edn. Harper Collins, New York, USA.

Huheey, J.E., Keiter, E.A. and Keiter, R.L. (1993) *Inorganic Chemistry – Principles of Structure and Reactivity* 4th edn. Benjamin Cummings.

Sections E, I and J	Fox, M.-A. and Whitesell, J.K. (1997) *Organic Chemistry*, 2nd edn. Jones and Bartlett, London, UK.
Section M	Bailey, P. D. (1992) *An Introduction to Peptide Chemistry*. John Wiley & Sons, Chichester, UK.
	Blackburn, G.M. and Gait, M.J. (Eds), (1996) *Nucleic Acids in Chemistry and Biology*, 2nd edn. Oxford University Press, Oxford, UK.
	Jones, J. (1992) *Amino Acid and Peptide Synthesis*. Oxford Chemistry Primers No. 7, Oxford University Press, Oxford, UK.
	Saenger, W. (1983) *Principles of Nucleic Acid Structure*. Springer-Verlag, New York, USA.
Sections N, O and P	Chang, R. (1981) *Physical Chemistry with Applications to Biological Systems*, 2nd edn. Collier-Macmillan, London, UK.
Section Q	Banwell, C.N. and McCash, E.M. (1994) *Fundamentals of Molecular Spectroscopy*, 4th edn. McGraw-Hill, London, UK.
	Breitmaier, E. (1993) *Structure Elucidation by NMR in Organic Chemistry*. John Wiley & Sons, Chichester, UK.
	Campbell, I.D. and Dwek, R.A. (1984) *Biological Spectroscopy*. Benjamin Cummings, Menlo Park California, USA.
	De Hoffmann, E., Charette, J. and Stroobant, V. (1996) *Mass Spectrometry – Principles and Applications*. John Wiley & Sons, Chichester, UK.
	Evans, J.N.S. (1995) *Biomolecular Spectroscopy*. Oxford University Press, Oxford, UK.
	Field, L.D., Sternhell, S. and Kalman, J.R. (1995) *Organic Structures from Spectra*, 2nd edn. John Wiley & Sons, Chichester, UK.
	Harris, D.A. and Bashford, C.L., (Eds), (1987) *Spectrophotometry and Spectrofluorimetry – a Practical Approach*. IRL Press, Oxford, UK.
	Hollas, J.M. (1996) *Modern Spectroscopy*, 3rd edn. John Wiley & Sons, Chichester, UK.
	Hore, P. J. (1995) *Nuclear Magnetic Resonance*. Oxford Chemistry Primers No. 32, Oxford University Press, Oxford, UK.
	Markley, J.L. and Opella, S.J. (Eds), (1997) *Biological NMR Spectroscopy*. Oxford University Press, Oxford, UK.
	Pavia, D.L., Lampman, G.M., and Kriz, G.S. (1996) *Introduction to Spectroscopy*, 2nd edn. Saunders Golden Sunburst Series, Fort Worth, USA.
	Silverstein, R.M. and Webster, F.X.(1998) *Spectrometric Identification of Organic Compounds*, 6th edn. John Wiley & Sons, Chichester, UK.

INDEX